# From My Experience

Other books by Louis Bromfield:

*The Farm*

*Pleasant Valley*

*Malabar Farm*

# From My Experience

## The Pleasures and Miseries of Life On a Farm

## LOUIS BROMFIELD

*with an Introduction by Janet Flanner (Genêt)*

The Wooster Book Company
Wooster • Ohio
2000

The Wooster Book Company • Wooster Ohio • 44691

Copyright © 1955 Louis Bromfield
Copyright renewed © 1983 Hope Bromfield Stevens
Introduction Copyright © 1999 William Murray & The Wooster Book
   Company – Originally commissioned by *The New Yorker*.

ISBN 1-888683-55-4

*Library of Congress Cataloging-In-Publication Data*

Bromfield, Louis, 1896 – 1956.
From my experience : the pleasures and miseries of life on a farm / Louis Bromfield ;
   with an introduction by Janet Flanner.
            p.               cm.
Originally published: New York: Harper Brothers, 1955.
ISBN 1-888683-55-4 (pbk.: acid-free paper)
   1. Bromfield, Louis, 1986–1956—Homes and haunts—Ohio.
   2. Authors, American—20th century Biography.
   3. Agriculture—Ohio—History—20th century.
   4. Farm life—Ohio—History—20th century.
   5. Farmers—Ohio Biography.
I. Title
PS3503.R66Z464            1999
813'.52.dc21                                          99-35558
   [B]                                                CIP

# From My Experience

# CONTENTS

## A FINE VERBAL CLIMATE
## An Appreciation of Louis Bromfield

Louis Bromfield is, fundamentally, an American provincial who temporarily transplanted himself onto the soils of France, India, England, Ritz bars in any land, and the skiing slopes around Saint Moritz. During those fourteen years that he lived outside his native Ohio, his conversation became an increasingly lively personal mixture of excellent old-fashioned French, the latest Broadway slang, dukes' and duchesses' first names, discussions on farm manures, Paris politics, Bombay palace intrigues, modern European painting, GOLDEN BANTAM corn, the operas of Richard Strauss, the best way to cook hamburgers, gossip about Hollywood movie stars, gossip about his French and American village neighbors, gossip about history over the last two- or three-thousand years, and talk about people, alive or dead, in general. Bromfield is at his best and most golden in conversation, the way a yellow-haired woman is at her best in the sunshine.

The one thing which Bromfield probably never talked of in those fourteen foreign years were the ten novels, three books of short stories, and the three plays he was writing. In one respect at least, Bromfield is the perfect novelist. He constantly writes novels and never talks about them. As a writer his output has been prodigious. He first started writing at the age of twenty-five. Except for the tiny lines around his eyes he now looks forty-five, but is really fifty-five and has just finished writing his nineteenth novel. As a matter of fact, he has really written twenty-four novels and his so-called first, *The Green Bay Tree*, was really his fifth because behind it lay four preliminary firsts which he wrote to learn how and to tear up.

The gift that makes Bromfield click as a human being, writer, talker, and farmer is his excessive energy. He has four times too much energy and no more than he needs to give one-hundred percent to being each of the four things he is. Because he is a solid, sandy, long-legged country boy, he has been able instinctively to rotate himself, the way a farmer rotates his crops. Whereas a city fellow, in Bromfield's shoes, would probably only have gone airily to seed. Out of what looks to outsiders like the complicated bedlam of Bromfield's life, he has always created a coherent pattern which he was able to follow because he is a dynamo, a realist (with a few romantic flowers in his buttonhole), and an efficient eccentric who thinks the beaten path is as dull as it is unworthy. When he entered Cornell Agriculture at the age of sixteen, he planned to become a farmer and to take each of his four university years in four different colleges. Instead, he polished off only Cornell and Columbia, and an honorary degree. He volunteered as an ambulance driver attached to the French Army, did the last two years of the First World War all the way from the Swiss border to the North Sea, learned to drink pinard, to speak French, to love France; came back to New York, got a job on the Associated Press, and in the next few years worked for a play producer on Broadway, for a publishing house, was on the original staff of *Time Magazine*, was book critic for *The Bookman*, became foreign editor for *Musical America*, a young husband, and went back to Europe to live.

The basis for Bromfield's life and work is people. He collects people (and notes their value) the way some men collect stamps. There must be seven- or eight-hundred human beings of many races scattered around the globe whom Bromfield knows well enough to have learned their life story at one sitting—well enough for him instantly to remember their faces, names, and the history of their hearts when he next meets them in a Basque bar in California, thousands of miles and years away from their first meeting which might have been in an Arab restaurant in Suez, at an Italian opera in Austria, or at a smorgasbord in Bloomsbury. And there must be at least one-hundred people whom he knows positively intimately, becoming an expert on their private lives and characters. What

Bromfield is really interested in, astounded by, and most patient with, is as old as the hills—human nature. With ideas, which always have a date-line, he is sometimes contentious, hasty. Probably the reason why American intellectuals have obstinately undervalued Bromfield as a nov-elist is because today's intelligentsia have been limitedly interested only in a hero's brain rather than all of his organs and what makes him and them turn out the way they do, according to when and how and where they were living and loving. Because of his passion for all types of people, Bromfield is probably as close to a twentieth century Balzac as the United States possesses and if he has come no closer, it's because contemporary souls are relatively poor stuff and cafe society contains shallower fools in our time than it honored even in Balzac's. Anyhow, making people's pri-vate lives come alive, on paper, hasn't been especially appreciated in the 1920s and '30s in which millions of Americans have led the only kind of life they had in public. For the writer Bromfield, France was a fine verbal climate. In the years near Paris he carefully learned from the French what they finally weren't able to use to save themselves—realism, or the adult talent of seeing things as they are.

In France, Bromfield also learned the value of being born and remain-ing a provincial. A French provincial is what we apologetically call a Midwesterner. The Parisians may have been crazy about themselves but the rest of the French were not. The rest of the French were sturdy provin-cials—people who knew the fundamentals of life didn't change like the styles in Parisienne's skirts. This country's common sense theory was right up Bromfield's Ohio alley. His people had come out to Mansfield from New England and Maryland in the early 1700s and his grandfather had farmed his own Buckeye land for sixty years. In his never-fed-up appetite for meeting, watching, and talking to human beings, Bromfield may have met more maharajahs, courtesans, White Russian taxi drivers, decayed nobility, deck stewards, French peasants, milords, millionaires, newspaper men, Belgian motorcar salesmen, nightclub singers, bums, and honest men and women of varied colors than most Ohio boys. But neither meeting them nor turning some of them into ink afterward has changed his original shape. He's a country man who happens to plow with his pen.

His gardens at Senlis, the Gothic town where he made his home, were famous for their flowers and vegetables. He has the green thumb and can make anything grow, principally because he will sweat hours in the sun with a hoe. An ambassador invited to the Bromfields for lunch mistook Bromfield, in the GOLDEN BANTAM corn patch, for the hired man. Miss Dolly Wilde, witty niece of her witty uncle Oscar Wilde, complained that Louis's flower garden had too many blossoms to be refined. His six-foot delphinium, his man-high canterbury bells, his bushes of roses maddened the French who expected an American to be better than they only in the matter of bathrooms.

Out of his garden grew one of the calmer friendships of Bromfield's feverish life—his friendship with the late American novelist, Mrs. Edith Wharton, his neighbor in the French countryside and also a great gardener. They used to discuss the technique of de Maupassant and tulips together. Bromfield is an omnivorous reader and rememberer. His other important literary friendship, though it included more loud laughing and joking, was with Miss Gertrude Stein whom he is devoted to. She was a steady habitué of his famous alfresco Sunday lunches when thirty or forty friends, many of them invited, would turn up at Senlis for cold cuts and strawberry shortcake served at tables all over the flower-bordered lawn. If all the people Bromfield has paid food and drinks for were suddenly to return the hospitality, he'd have to be a Gargantua to get away with it all.

Miss Stein, who is a Republican, thought Bromfield knew nothing about politics because he was a Democrat but conceded that he was anyhow the earliest and best Jeffersonian of her acquaintance. Sometimes he thinks he'd like to go into politics, be a rural senator or something. He's not avid for honors. Though he won the Pulitzer Prize for his novel *Early Autumn*, he likes three others better—*The Strange Case of Miss Annie Spragg*, *The Rains Came*, and *The Farm*. Who's Who states that Bromfield was awarded the *Croix de Guerre* but none of his friends ever hear him mention it. The only prize he ever crowed over was a second prize in skiing he won in a Saint Moritz tournament. He thought it a vindication of the Ohio skiing school in which, as a boy, he learned to ski on barrel staves.

Bromfield writes his books in pencil, longhand. He has such concentration that he can come in from working in his fields, go to his desk and finish a sentence he started the day before. He loves music, jazz and classic, has a superb record collection of Mozart and Strauss; *Don Giovanni* and *Der Rosenkavalier* being two of his favorites. Though he lived in

France for years, he's no pretentious wine taster; he's a highball man. He's certainly no dude. If Bromfield remembers to wet and comb his hair when he dresses, he has a juvenile air of having reached his limit in male swank. His favorite sports are skiing and swimming, and he did some tiger hunting in India. His best pastime, though, is conversation. Anybody who did not like Louis Bromfield and listened to him talk for an hour would at any rate understand why other people are charmed into loving him.

Bromfield is the father of three daughters, one dignified, and two that are lively and nimble. His wife is the former Miss Mary Appleton Wood, an exceptional humorous member of a distinguished old Park Avenue family.

The Bromfields live in Ohio on his six-hundred acre investment, Malabar Farm. It is beautiful, but there's no nonsense about it. It's a series of five-year plans of Bromfield's, with beef, pigs, lambs, chickens, bees, fodder, and food as its basis. The government is using part of it as an agricultural experiment station. Bromfield works with the local farmers on the community's cooperative projects, is doing a swell, conscientious, successful, responsible job. Bromfield's favorite Frenchman was a sensible little factory worker named Picquet, who, on less than a dollar a day and with one acre of fertile, well-planted ground, was able to find security, nourishment, philosophy, and peace—while it lasted—for him and his family. Because Bromfield has seen so many different lands, he is now more a country man than ever. When he turned his first spadeful on his new Ohio farm acres, it marked the return of the native.

—JANET FLANNER (Genêt)
for *The New Yorker*

# From My Experience

## APOLOGIA

IF AT TIMES IN THIS BOOK THE TONE OF WRITING APPEARS TO BE UNDULY controversial I attribute this to long contact with many of the closed minds and the unimaginative mentalities with which agriculture, like any other science, is afflicted. The writer is by nature an amiable and kindly person, very gregarious and fond of people and of conversation, of argument and even of controversy, and the exploration of other minds where there is anything to explore. But he has had some startling experiences with the closed mind which will not accept what it sees and knows but takes refuge sometimes in wild and fantastic speculation and sometimes feeble diversions to find reasons to deny or discredit a fact that may prove unwelcome or embarrassing. In no sense is this true of such men as William Albrecht, Firman Bear, Sir Albert Howard, the late Ed Babcock, and the late Washington Carver and countless others who have made great contributions to the vast and complex field of agriculture. I might also add that the usually good-natured and easygoing author of this book has a trigger temper which can, like some explosives, be put into action only by certain elements in certain combinations and under certain conditions. The trigger factor in the case of this book is lack of imagination and the closed mind which believes, sometimes with sincerity, sometimes out of mere laziness, that everything has been discovered and that nothing more can be learned about the universe. It is possible that to some this book may seem lacking in organization, but this is not so. In the mind of the author at least, the material has been put together with cunning, much as the use of bait leads an animal into a trap. Some of the chapters require hard concentration on the techniques of farming, some are concerned with philosophical reflection, and others may seem light to the point of triviality.

I have long been aware that much reading on the subject of agriculture is extremely heavy and stuffy and that in many cases it consists largely of papers written by one scientist *at* another and set forth in language so pompous and so technical that the effect becomes lethal. While I do not believe that this record of the Pleasures and Miseries of country living will entertain the reader from beginning to end, the passages requiring extra-close attention exist in spots between stretches of lighter and easier reading. In any case, daily life on a living farm with its countless facets, its daily crises, its seasonal changes, resembles very closely the general pattern of this book—a record of ups and downs, which must be taken, if at all, as they come.

## FIFTEEN YEARS AFTER
(By Way of Introduction)

F IFTEEN YEARS HAVE PASSED SINCE THE SNOWY WINTER DAY WHEN I turned the corner of the road into Pleasant Valley and said to myself, "This is the place." I had come back after twenty-five years of living in the world, to my own country, to the valley I had known as a boy. I knew the country in the marrow of my bones; I knew it even in the recurring dreams which happened in strange countries here and there over half the world. I knew the marshes and the hills, the thick hardwood forests, the wide fields, and the beautiful hills behind which lay one lovely small valley after another, each a new, a rich, mysterious self-contained world on its own.

I was sick of the troubles, the follies, and the squabbles of the Europe which I had known and loved for so long. I wanted peace and I wanted roots for the rest of my life.[1]

When I saw the valley again after twenty-five years it was under deep snow and the farms I bought were under deep snow. I have recounted the whole story in detail in *Pleasant Valley*, while the first impressions were still fresh in my mind, and I am glad that I did this for I can read it all and know now how little I understood the changes that occurred while I was

---

[1] I cannot say that with regard to the troubles, the follies, and the squabbles I found much change or relief. The record of my own country in these times with its politics, its meddling in the affairs of other nations, its spasmodic Utopianism, its militarism, its saber rattling, its attempts to dominate the world and dictate the policies of other nations, has been no record in which to take pride or to justify a sense of superiority in any American.

away and how little I foresaw or understood what lay ahead and how nearly all the values of my life were to be changed and enriched. What I saw then, I saw through a haze of nostalgia, with homesick eyes. What I was seeking in part at least was something that was already on its way out of American life.

On that first snowy evening when I knocked at the door of the farmhouse which stood where the Big House now stands, I was, like many a man on the verge of middle age, knocking at the door of my long-gone boyhood. Tired and a little sick in spirit, I wanted to go back and, like many a foolish person, perhaps like all of us, I thought or hoped that going back was possible. I was sick even of writing novels and stories, although they had brought me considerable fortune and fame in nearly every country in the world. All fiction, save perhaps such books as *War and Peace*, which is more history than fiction, seemed to me at last to be without consequence and even trivial in contrast to all that was going on in the world about me. I knew and partly understood—better than most, I think—what was going on and what was ahead, certainly far better than the great majority of Americans, because I had lived for nearly a generation at the very midst of the turmoil and the decay which ended finally with the humiliation of Munich and the Second World War.

Most of the fellow countrymen to whom I talked on my return seemed almost childish in their naiveté and their lack of understanding concerning the significance of what was happening in Europe and in the world—just as many of them today seem childish in their refusal to face a world which is utterly changed, a world in which Soviet Russia and Red China and the awakening of Asia and the decadence of Europe and the end of the colonial empires are all simple facts which cannot be wished nor laughed off nor evaded. Peace and decency can only come by and through recognition of such facts and a recognition above all that the old world which many of us perhaps found agreeable enough is not coming back.

Those of us who lived in Europe and Asia between the first two world wars knew and understood pretty well both Nazism and Fascism and we knew too all about Communism in all its manifestations, and knew that there was nothing to choose between the doctrines; one was merely a perversion of an unnatural political and economic philosophy and vice versa. They were both derived from the philosophy of a sick and psychopathic German called Marx. The forces behind them were actually the forces of a world revolution which manifested itself in countless ways, not the least of which was the slow death of the already obsolete colonial empires. These doctrines were created not merely by a philosophy, wishful thinking, or even fanaticism: they were created by immense economic and political stresses and strains which involved such things as overpopulation, shortages of foods and raw materials, abysmal living standards, and restricted markets. In short, Nazism and Communism were merely the political manifestations of profound economic ills. They did not in themselves cause war and revolution. They were basically the outward political manifestations of much more real and deep-rooted evils. The guilt for the war could not be fixed on any one nation or group of nations. Every nation in one degree or another shared responsibility. The responsibility still remains. We in this country have solved nothing nor made any great contribution toward creating and maintaining peace. The Utopians and the militarists have indeed merely accomplished the contrary.

Fifteen years after I hoped bleakly to escape from all the evils I knew so well at first hand, I have discovered bleakly that there is nothing superior about my own people and that they do not have any special wisdom or vision. We have merely been more fortunate than other peoples. We are generous because we can afford to be generous. We are perhaps openhearted because we are still a young people, but we still understand very little about the evils of the world or how they can be cured or at least modified. We lack almost entirely the capacity of putting ourselves in the place of other peoples and the knowledge of the average citizen concerning the life and the circumstances of other nations and peoples is primitive, frequently enough even among those who occupy high places in our government.

When I came back to the valley on that snowy morning fifteen years ago, I was trying to escape the evils of the past and the weariness of Europe. Somehow in the misty recesses of my mind there existed a happy image of this valley in which I had spent a happy and complete boyhood—the image of a valley shut away, immaculate and inviolate, self-sustained and complete and peaceful. It was an image in which the sun was always shining. It was not of course an image born of the mind and the intellect but one born of the emotions. At that time I did not fully understand what it was that moved me so powerfully. I have only come to understand it through the gradual corrosion of disillusionment and pessimism, and since then I have found something which I did not then know existed, something which I did not even know I was seeking ... something which is difficult or impossible to describe save that it is a combination perhaps of reality and truth and of values which were unsuspected. All this is closely related to the earth, the sky, to animals and growing plants and trees and my fellow men. All this is of course of immense importance to me personally; it is important to others only as a bit, a fragment, of human experience, that element which perhaps more than any other sets men apart from animals who, so far as we know, are not capable of reflection or philosophy.

And so in the light of all this, the writing of fiction, unless it was merely a story to divert a tired world or provide relaxation for it, came presently to seem silly. It still does, no matter how pompous, how pretentious, how self-important, how cult-ridden the writer or the product. In this age fiction writing is simply a way of making a living and for my money not a very satisfactory or even self-respecting one. There are better and more satisfying things to do. One degree sillier are the writings of those who write importantly about novels. Once when a person said to me, "Oh, I never read novels!" I was inclined to regard him with snobbish condescension as a Philistine. Now I am not so sure.

So when the door opened in the Anson house more than fifteen years ago and the familiar childhood smell of a farm kitchen came to me, I was aware of a sudden delight as if in reality I had stepped from the recurrent weariness and disillusionment back into the realities of my boyhood. The smell was one in which there was blended the odor of woodsmoke, of apple butter, of roasting pork sausage, of pancakes, of spices. Mrs. Anson, workworn and no longer young, her arthritic hands wrapped in her apron, stood in the doorway with her back to the gaslight of the big kitchen walled with hand-cut stone from the low cliffs behind the house. With a puzzled smile lighted by some ancient memory and recognition, she invited me in, and brought me hot coffee and spiced cookies. When I told her my name she remembered me dimly as a small boy who used to camp on the adjoining Douglass Place which belonged to her cousins, and who fished in Switzer's Run that flowed through the valley below the house.

When I asked her if the place was for sale, she answered that she believed not but that we might talk to her husband who was out in the barn milking. And so she put on a shawl and we walked in the blue winter twilight across the squeaking snow to the barn where old Mr. Anson sat on a three-legged stool drawing milk from a Guernsey cow. And again at the barn door the smell of the warm stable, the granary, the steaming manure came to me in the form of the perfumes of Araby. For the moment I again renounced all the world in which I had lived most of my life and escaped from it back into the past … into the world of horses and buggies, of muddy roads, of church suppers, of everything that was the America of my childhood. I who had been everywhere and known the world and "all the answers" wanted only to come back into this world.

The Ansons could not make up their minds whether they wanted to sell until they talked to their children, and so after another cup of coffee in the big stone kitchen I set off down the winding road past the cottage where Ceeley Rose had poisoned her parents and brothers, across the little bridge, and finally to the highway. It was a *blue* winter night with that peculiar quality of blue in the sky and in the air itself which one finds in our part of the world and only a few other places; places like northern France and the Castilian plain about Madrid. The air seems to become luminous like the unreal blue of skies on the cyclorama of a theater. I was happy. I was hopeful that I had really escaped and had come home.

That night the whole valley was covered by snow and the little creek fringed with ice. What lay beneath the snow I could not see. I was only to discover it when the spring came with a rush of green and wildflowers.

When the snow was gone, I discovered that the valley of my childhood was no longer there. Something had happened to it. It had been ravaged by time and by the cruel and careless treatment of the land. As a small boy I had never noticed that these once small, lovely, rich valleys throughout our countryside had already begun to change, growing a little more gullied and bare with each year, or that the pastures grew thinner and more weedy and the ears of corn a little smaller each season. When the snow was gone, I began to understand what had happened to it … perhaps because for so long I had lived in the rich green country of northern France where the land is loved and where it is respected and cherished almost with avarice, perhaps out of grim necessity, but none the less respected and even held precious. It was clear that no such thing had happened in the valley. Some of the farms which lay below and around the Anson Place no longer raised crops at all. They had been rented out to year-by-year tenants or to neighbors who took everything off them until they would no longer grow anything. The houses were occupied by industrial workers who spent their days in the city factories.

Except where starving sheep on wasted farms were turned loose in the woods to find meager living as best they could off the seedlings and the sun-starved vegetation, the forests and the marshes were still the same. Here and there a woodlot had been brutally murdered by some fly-by-night timber speculator, the trees sold perhaps to raise money for interest on the mortgage of some dying farm; but otherwise the growth of the wild grasses, the trees, the briars provided evidence of the original and fundamental richness of the whole countryside. Wherever there was desolation and sterility it had been created by man, by ignorance, by greed, and by a strange belief inherent in early generations of American farmers that their land owed them a living.

When I look back now, the vague and visionary idea I had in returning home seems ludicrous and a little pathetic. Somehow the picture I saw of the future was one in which vaguely there were blended the carefree happiness of my boyhood and the life in a great house in the countryside of England which, in the great days, was perhaps the best and most civilized life man has ever known. If in the dream there was any other element, it was that of security; I wanted a place which, again vaguely, would be like the medieval fortress-manor of France where a whole community once found security and self-sufficiency. In a troubled world I wanted a place which, if necessary, could withstand a siege and where, if necessary, one could get out the rifle and shotgun for defense.

Today, fifteen years later, we at Malabar have not achieved these romantic dreams nor have I won the escape into the boyhood past which brought about the decision to return. A return to the past can never be accomplished and the sense of fortified isolation and security is no longer possible in the world of automobiles, of radios, of telephones and airplanes. One must live with one's times and those who understand this and make the proper adjustments and concessions and compromises are the happy ones. In the end I did not find at all what I was seeking on that snowy night; I found something much better ... a whole new life, a useful life, and one in which I have been able to make a contribution which may not be forgotten overnight and with the first funeral wreath, like most of the writing of our day, but one which will go on and on. And I managed to find and to create, not the unreal almost fictional life for which I hoped, but a tangible world of great and insistent reality, made up of such things as houses, and ponds, fertile soils, a beautiful and rich landscape, and the friendship and perhaps the respect of my fellow men and fellow farmers. The people who come to the Big House are not the fashionable, the rich, the famous, the wits, the intellectuals (although there is a sprinkling of all these), but plain people and farmers and cattlemen from all parts of the world.

Perhaps it will turn out that I have left behind some contributions not only to the science of agriculture, which is the only profession in the world which encompasses *all* sciences and *all* the laws of the universe, but to the realm of human philosophy as well. None of this could I have done within the shallow world of a writer living as most writers live. Without implying in any way a comparison or any conceit, I am sure that Tolstoy understood all this on his estate at Yasna Polyana, Voltaire at Ferney, Virgil on his Tuscan farm; indeed most writers since the time of Hesiod who felt sooner or later the illusion and the futility of fancy words and sought some sturdier and earthier satisfaction. Flaubert was neither a happy man nor a complete one, nor was Turgeniev or Henry James. One has only to read their letters or their diaries and sometimes their stories and poems to discover a shadowy sense of impotence and inadequacy.

The *complete* man is a rarity. Leonardo was one and Michelangelo and Shakespeare and Balzac. They lived; they brawled; they had roots; they were immoral: they had vices as well as virtues; they were totally lacking in preciosity and the pale, moldy qualities of the poseur or the seeker after publicity and sensation. The *complete* man is a happy man, even in misery and tragedy, because he has always an inner awareness that he has lived a *complete* existence, in vice and virtue, in success and failure, in satisfaction and disappointment, in distinction and vulgarity. Not only is he complete, he is much more; he *is* a man.

The older I grow the more I become aware wistfully of that goal of completeness. It is not something that can be attained by wishing or even by plotting and determination. The man who sets out deliberately to be a complete man defeats himself, for from the beginning he is of necessity self-conscious, contriving, and calculating. He becomes the fake, the poseur, the phony. Some attempt to turn their own inadequacies into a defense by affecting a sense of snobbery or superiority. In this sense a writer like Henry James is pathetic. So are many writers of our own times with their lacy preciosities, their affectations, their pomp and pretensions, their fundamental shallowness and decadence.

But enough of all that. If such thoughts have any importance, it is an importance which primarily touches only the individual and the individual only in relation to the satisfactions which he may win from a life which is all too short if one considers the glories, the complexities, the mysteries, the fascination, of the world and the universe in which we live. Certainly one of the happiest of men is the good farmer who lives close to the storm and the forest, the drought and the hail, who knows and understands well his kinfolk the beasts and the birds, whose whole life is determined by the realities, whose sense of beauty and poetry is born of the earth, whose satisfactions, whether in love or the production of a broad rich field, are direct and fundamental, vigorous, simple, profound, and deeply satisfying. Even the act of begetting his offspring has about it a vigor, a force, a directness in which there is at once the violence of the storm and the gentleness of a young willow against the spring sky.

I was born of farming and land-owning people and have never been for long away from a base with roots in the earth, despite the fact that the fortune and circumstances of life have from time to time brought me into the most worldly and sophisticated and fashionable of circles in a score of countries and capitals and frequently among the great, the famous, and those in the world of politics, who have made the history of our times. Still I know of no intellectual satisfaction greater than that of talking to a good intelligent farmer or livestock breeder who, instinctively perhaps, knows what many less fortunate men endeavor most of their lives in vain to learn from books, or the satisfaction of seeing a whole landscape, a whole small world change from a half-desert into a rich, ordered green valley inhabited by happy people, secure and prosperous, who each day create and add a little more to the world in which they live, who each season see their valley grow richer and more beautiful, who are aware alike of the beauty of the deer coming down to the ponds in the evening and the mystery and magnificence of a prize-winning potato or stalk of celery, who recognize alike the beauty of a field with a rich crop in which there are no "poor spots", and the beauty of a fine sow and her litter.

These are, it seems to me, among the people who belong; the fortunate ones who know and have always known whither they were bound, from the first hour of consciousness and memory to the peace of falling asleep for the last time never to waken again, to fall asleep in that tranquility born only of satisfaction and fullness and completion. In a sense they know the whole peace of the eternal creator. They have built and left behind them achievements in stone, in thought, in good black earth, in a painting upon a wall, or some discovery which has helped forward a little on the long and difficult path their fellow men and those who come after them. It is these individuals who belong and who need not trouble themselves about an afterlife, for at the end there is no terror of what is to follow and no reluctance to fall asleep forever, since it is the direct and the natural thing for them to do. They are at once and in essence humble and simple people, no matter what fame they achieve or what admiration they have won from their fellow men. Because they belong and so have found their place and their adjustment in the intricate and complex pattern of the universe they find even the tragedies and the suffering of their own lives only a part of a general vast pattern. They are, perhaps, more sharply aware of the significance of tragedy and suffering than less limited and egotistical folk; that very awareness and significance blunt the sharp edge of suffering and presently the pain wears away, leaving only peace and even perhaps a strange sense of beauty and richness.

The mysteries of the human mind are certainly fascinating but more limited and, for me at least, less important and less interesting than the cosmic mysteries which take place within a cubic foot of rich productive soil; for essentially these "mysteries" of the human mind are merely a part of an infinitely greater and more intricate and complex mystery which utterly baffles all of us, even the wisest and most learned. One of the great errors of our time, and one which has brought us in our time much misery, is the attribution of an overweening and disproportionate importance to man and his mind. Man himself, as a physical machine, as a mechanistic functional and living organism, is indeed marvelous as is every part of the universe; but his ego and self-importance, in our time, are given a distorted, decadent, and tragicomic importance.

Man is merely a part of the universe, and not a very great part, which happens to be fortunate principally in having evolved such traits and powers as consciousness, reflection, logic, and thought. The wise and happy man is the one who finds himself in adjustment to this truth, who never needs, in moments of disillusionment and despair, to cut himself down to size, because it has never occurred to him, in the beginning or at any time, to inflate his own importance, whether through ignorance, morbidity, egotism, or undergoing psychoanalysis (which is merely another name for one of the age-old manifestations of brooding impotence and frustration of the incomplete man).

It is only later in life, in the midst of what is still a somewhat turbulent and certainly a varied existence, that any full understanding and satisfaction of this sense of *belonging*, of being a small and relatively unimportant part of something vast but infinitely friendly, has begun to come to me. It is only now that I have come to understand that from earliest childhood, this passion to *belong*, to lose one's self in the whole pattern of life, was the strong and overwhelming force that unconsciously has directed every thought, every act, every motive of my existence.

This urge is by no means an uncommon one and is perhaps shared in one degree or another by every man or woman of real intelligence, although many fail to recognize it for what it is or to understand it. Certainly it is the profoundly motivating pattern in the life of a man like Albert Schweitzer, just as it is of many a humbler thinker and teacher. It is indeed the very force of every good and real teacher; it is that element which distinguishes the creator from the desiccated savant, that distinguished Christ from Socrates, that makes all formal education seem futile and useless unless it is *understood* and *used*. Mere *learnedness* becomes insignificant and even useless when the fire is absent. This rarity of understanding is the reason why throughout the ages there has always been a cry that there are not enough teachers, and why all the knowledge and all the formal training in the world produced by all the universities and colleges since the beginning of time cannot produce one more real teacher. I am afraid it is true that teachers, like good cattlemen and good farmers and good engineers and good cooks, are born and not made. So essentially are most really good people.

The egotistical, the resentful, the psychologically maimed can never really belong. Perhaps in the end the great failure of the Communist doctrine will be registered in history as a failure to produce great teachers, or any teachers at all, or indeed any complete men who really belonged. The typical leader has been, almost without exception, the psychologically

deformed, the egotist, the man hungry for power but not for understanding, altogether the man confused and limited by the stupidities of
materialism and the domination and tyranny and terrors that spring
from and surround his own ego. Few men in human history have had
such power as Stalin, but what honest and intelligent man would ever
wish to have changed places with him? It is probable that he was a most
limited man who never knew even the shadow of that satisfaction which
comes to the complete man.

And now, dear reader, if you have survived these philosophical ram-
blings we shall get on to the business at hand ... the business of drying
hay, of finding means for making more money for the farmer, of helping
to feed hungry people, and in general the satisfactions of life in the country. But do not overlook the fact that all of these things—real, finite, and
even perhaps materialistic—are all very much a part of the pattern in the
life of at least one busy and happy man, even though he is a long way
from being a complete one, just as they have become in these times more
and more a part of the pattern in the lives of countless middle-aged and
elderly businessmen, industrial workers, engineers, clerks, retired
mechanics, and have been part of the pattern in the lives of a number of
Presidents, from Washington and Jefferson to General Eisenhower ...
men who have discovered sooner or later that somehow farming, and
especially good and intelligent farming, comes closer to providing a key
to adjustment and understanding of the universe than all the algebraic
formulas of Albert Einstein.

Considering the general insignificance and unimportance of man, the
pleasures of agriculture are perhaps more real and gratifying than the
pleasures and even the excesses of the purely mathematical mind (which
are certainly not pleasures to be underestimated). If the pleasures of
mathematical debauchery or orgies in physics are to be treated as limited, it is only because they are denied the great mass of humanity and
because all too often they induce and create the deformities and limitations of the incomplete man. Rousseau was in many respects a fool and
at times a humbug and a liar but he had something in his conception of
the natural man, something in which even the great, wise, and cynical
mind of Voltaire, an infinitely more intellectual and sophisticated man,
found a perpetual source of envy.

The greatest creative and intellectual vice of our times, and a factor which causes increasing distress and even tragedy, is the overspecialization which man has partly chosen and which has been partly forced upon him by the shrinking of the world, by the incredible speeding up of daily life, and the materialistic impact of technological development upon our daily existence. The super-specialist tends to become not only an incomplete man but a deformed one. What perhaps limits forever the super-specialist or the "pure" intellectual stewing in his own juice is the fact that in his intensity and in the sharp limitations of his narrow field, he loses frequently the power to understand or even to conceive the principle of the complete man; flesh, appetites, weakness, follies and all; and so he loses in time all understanding in the true sense of the universe, or even of the small world which surrounds him.

## A HYMN TO HAWGS

WHOEVER HAS REALLY LOOKED INTO THE EYE OF A SHREWD OLD SOW should feel humility. It is a bright clear eye, more like the eye of a human than the eye of any other animal. It looks at you quite directly, even with what might be described as a piercing gaze. The look sizes you up, appraises you, and leaves you presently with the impression that the old sow has indeed a very low opinion of you; an opinion tempered by scorn and contempt and perhaps even a little animosity. Clearly she does not think that you amount to very much, and that given a difficult situation she could cope with it far better than you could do. It is as if she said, "You think you can shut me up and confine me. Well, that's what you think! Ha! Ha! and again Ha! Ha!" And any farmer knows what she is thinking … that if she really wants to get out she'll find a way. Sometimes I have a feeling that she is thinking, "You think you know how to manage me and bring up my litters with all your disinfectants, your heat lamps, your violet rays, your antibiotics, your supplements, your inoculations, your vaccinations. Just let me alone and give me my freedom and I won't have any troubles, nor will the pigs I feed."

Yet she can be friendly too and even understanding, as any farmer knows when a seven-hundred-pound sow comes running to have her back scratched. You don't have to scratch her back a dozen or a score of times. Once is enough and she gets the idea and after that she'll always be on hand when you enter the field. She may be using you and undoubtedly knows that she is, and there *may* be some affection involved, but I doubt that she ever really loses her scorn for you. In some ways, about many basic and fundamental things, she knows both by instinct and intelligence quite a lot of things you don't know.

Animals at Malabar are likely to be well treated and most of them sooner or later come within the category of pets and sometimes very troublesome pets. Olle knows all his cows, what each one likes and doesn't like, in what order they prefer to enter the milk parlor, which cows have best friends, and which ones are bullies. He knows the bad temper of Inez, who just never could be milked without being tied down, and the comic ways of Lauren, who on more than one occasion was found shaking a barn cat in her mouth when the impudent cat attempted to walk across her parcel of feed. And he knows old Mary, whom he rides home from pasture to the milking parlor as if she were a horse. There are no ill-treated animals at Malabar, for the first consideration of employment in connection with any livestock is that the man must be able to imagine himself a cow, a pig, or a chicken, and so to know what would make that particular animal or bird happy, healthy, and comfortable. And so you get to know animals pretty well and the more you know them the greater respect you have both for them and for God and nature. The people who think all cows are exactly alike are merely stupid or ignorant; cows can sometimes be as individual as the people of any group and not infrequently a good deal more interesting and sympathetic.

But pets can be tiresome too, like the hand-raised lambs at Malabar who, growing up with the boxer dogs, came to believe they were dogs and not sheep at all, so that they even began to chase cars and provided the astonishing occasional spectacle of what appeared to be a sheep taking part in a dog fight. And when at a suitable age, an effort was made to send them off to live among their fellow sheep, they would not believe they were sheep and wanted no part of their kind, returning home to be with the dogs. All of which led us to the speculation as to what the lamb's mental image of himself might be—quite possibly the image of a large fawn-colored boxer dog with a black face and a protruding jaw.

The most troublesome of all farm pets at Malabar were the goats, which persecuted all of us and all visitors until they had to be sent away to Italian families who know better how to cope with them. The goats certainly arrived at a point where they were quite convinced that they were people and that the roofs of cars and the furniture on the terrace and porch and even the cook's bed had been designed and put there especially for them. If there is any animal more difficult to confine or to shut out than a pig, it is certainly a goat. Both are gifted with great powers of logic, reflection, judgment, deduction, and mockery, qualities which man has usually and often mistakenly reserved for himself. Perhaps this is the reason that both animals are closely associated with witchcraft and the unholy celebrations of Walpurgisnacht. It was not for nothing that in George Orwell's wonderful satire on Communism and Communists, he chose the pigs as the bureaucrat dictators of the barnyard and a pig as the symbol of Stalin.

Out of an excessively busy life, pigs have cost me more time than fishing or golf or any of the usual activities in which men are accustomed to find relaxation, amusement, and exasperation. When I am really busy, I dare not visit the pig lots, for inevitably I find myself standing there for indefinite amounts of time, watching the little pigs horsing about with their own games, listening to the conversation and gossip of the old sows, watching the pigs line up to feed, even on occasion watching a new litter arrive in this world to get up immediately on their feet, shake themselves, and run around to their outside source of supply in the new world which they have entered only a moment before. If I had ever doubted a pig's capacity to think things out, I lost it forever when I discovered one pig's trick of escaping through what was by every reasonable evidence a thoroughly pig-tight fence.

He was one of some two hundred half-grown pigs enclosed for feeding in a ten-acre lot. Every day, week after week, one pig got outside the fence and no one was ever able to discover how he accomplished this feat until late one afternoon as I stood in the far end of the field. I saw what I found difficult to believe. At the opposite end of the field I saw a pig actually *climbing* the fence.

The fence was a typical hog fence with graduated openings from those too small for even the smallest pig to negotiate at the very bottom to openings at the top that were reasonably large. But the designer of the fence had reckoned without this special pig who possessed advanced powers of reasoning. He had discovered that the higher the openings were from the ground the larger they were and he managed to climb high enough to reach an opening big enough for him to slip through. With the greatest dexterity he worked his way upward until he reached an opening of sufficient size and then slipped down on the opposite side of the fence.

I watched the whole operation, even to his scuttling across the road into a neighboring cornfield where he was just able to reach up and drag down the ears of corn. I must add that so great was my weakness for pigs and my admiration for this particular pig, that I never betrayed the fact that I had learned his secret. The end of the story was that eventually he ate himself out of his freedom, for each day he grew fatter, heavier, and larger in girth, until the day came when he was unable to find an opening to permit his escape or to reach the top of the fence without falling backward again into the prison of his hog lot.

And there was the mentally handicapped pig who never did learn to use the self-feeder and who, as a consequence, developed a psychopathic and extremely bad and aggressive disposition. For the benefit of the city-bred who may be reading, a self-feeder is constructed with a kind of trap door which the average pig, or indeed every other pig I have ever known save this one, learns very quickly to raise by putting his snout beneath it and forcing it upward. Once he has his snout inside the feeder, the trap door rests on his snout or rather his forehead until he has eaten his fill. Then, as he backs away, the trap door falls down again by gravity and the feeder is closed. Just why this trap door is necessary or why it is supposed to be useful I do not know, unless some bright fellow believed that it saved feed, although any farmer knows that any good healthy pig allows no feed to be wasted whether it is in the feeder or outside on the ground. The self-feeder, however unnecessary, has been forced into the consciousness of the average farmer, like many another dubious and expensive agricultural equipment product, by high-pressure advertising and the general pressure in colleges, industrial plants, and other places where new gadgets are constantly being thought up for farmers to buy. At this stage, I was still naive enough and inexperienced enough to believe that one could not raise hogs without a self-feeder, an idea which I long ago abandoned, for we at Malabar have never belonged to the school which invests in gadgets simply to make a hog lot or a building look "prettier."

In any case, we had in one lot of hogs this special backward pig who never learned to raise the trap door with his snout. He always took the lid between his teeth and raised it and always as he let loose of it the trap door fell shut again before he could get his snout under it to hold it open.

I have watched him try again and again countless times to perform the operation so rapidly that he could outwit the falling trap door. He never won and so in order to get his fill he turned into a bully. He would approach the self-feeder, take up a stand, and allow some innocent fellow pig to open the trap door with his snout. Then with a combination of a slashing and shouldering side movement, the backward pig would shove the innocent aside and thrust his snout into the opening to feed his fill. The odd thing was that he never learned, either by reasoning and logic, as did our fence-climbing friend, or simply by observing how the other pigs did it. He just never learned. But I think we need not be too critical or scornful of him; I have known a great many fellow men who suffered from the same handicap. He was the dumbest pig I have ever known and in fact the only "dumb" pig with whom I have ever been acquainted. He was also a bully, like many a human bully—like Mussolini or Molotov.

Much discussion and argument has been spent regarding the ability or inability of animals to communicate with each other, but nowadays I believe there are few people or even experts who any longer doubt that animals have means of communication which are quite beyond our understanding. Turn a bull into the far side of a field away from a herd of cows and heifers, with a hill or woodlot in between. All he need do—and he always does it once he is inside the gate—is to give one loud bellow and all the cows and heifers, tails high in the air, leaping and cavorting coquettishly regardless of age, will come running. What man has ever had the eloquence or the blarney to command so easily such a harem? Beside the bull Don Juan was merely a piker.

Or what about the cantankerous cow who manages to ride down a fence or find an opening and who, with a single bellow of peculiar volume and timbre, will bring all the rest of the herd to join her in breaking out?

Again and again I have seen one of my dogs soundlessly communicate to the others that there is a strange dog on the place or in a visiting car, which must be put in place or given the bum's rush. I have seen one of them tell the others that Lester the Cat was behind the greenhouse and in a vulnerable situation. A certain kind of bark surely communicates at least a certain variety or kind of information. But among all animals there is none which has such a variety of sounds designed obviously for the purposes of communication as the pig. Waste your time as I do, leaning on the fence, and you will see among the pigs everything from a director's meeting to a ladies' discussion of the attractions of the visiting boar. There are the sounds which a sow makes when she feels it is time to feed her young, and there are the frightful squeals of discontent from her pigs when she will not give in to their demands and lie on her side to let them feed. And there is what is distinctly the most ferocious name-calling that can occur during the equivalent of a hair-pulling match between two old sows. Pick up a small pig and he will let out the most hair-raising shrieks for help that will bring running not only his own mother but all the sows in the neighborhood. He is the most blatant of small frauds, for the minute you put him down again on the ground the shrieks stop instantly and are replaced by the pleasant "oink! oink!" noises of a contented small pig. Come unexpectedly upon an old sow and her litter in a swamp or a thicket and, especially if there are dogs along, she will give the signal of alarm, gathering all the litter around her where they form a circle with all their small noses pointed outward like a Macedonian phalanx both for the attack and defense, making at the same time the most threatening and blood-curdling sounds.

These sounds made among pigs are not merely signals or crude symbols of speech. Observe a gathering of sows when they are not eating or feeding their pigs but are merely having a good gossip. The hog wallow is clearly a favorite spot for woman's talk—the kind of talk you might hear on "Ladies' Night" in a Turkish bath. There is a whole variety of sounds, including not only pitch but intonation and the actual formation of given sounds. I once assisted at such a gathering among a group of sows who came to give advice at the birth of a new litter by one of their number.

The sow who was brought to bed was either taken unexpectedly in labor or was a plain fool, which is the more likely situation. She didn't choose or (like a woman giving birth in a taxi) failed to start for the shelter of the barn in time and consequently had her litter in the middle of the hog lot on a day of great heat and blazing sunshine, something no sensible sow would dream of doing. In any case, the family came rapidly, all ten of them, popping up on their wobbly feet and going around at once to feed at the waiting cafeteria. The spectacle attracted several other sows who stood about very clearly making comments of a disapproving nature. When the last of the pigs had entered this world, or, as the fashionable birth announcements in the French newspapers put it, "was given the light of day," the sow stood up and apparently either became aware of her foolishness in having her pigs in the midst of a hot field or heeded the reproaches and advice of the surrounding sows. She attempted to induce the pigs to follow her toward the shade of the nearby barn with no success, for the pigs were still so new to this world that, beyond the instinct of eating, they were without adjustment and clearly did not yet understand pig language. She made an astonishing variety of sounds, in which her friends joined ... sounds that were now cajoling, now scolding, now clearly meant as commands. But the tiny shaky pink pigs merely wobbled about aimlessly. Then, abandoning temporarily all hope of getting them into the barn, she relaxed for a time and held what was clearly a discussion of her labor pains and prenatal condition with her friends. There was the most astonishing variety of sounds and intonations in which the other sows, their heads close together, all joined. This continued for some time with a growing animation, for all the world like a group of women during a bridge game which has been interrupted by the more absorbing topic of whose labor pains were longest and worst. Then after a time she returned to her futile efforts to herd the tiny pigs into the shade.

Nothing came of her efforts and I went to get Bob and Carl to help rescue the pigs. They were of the Yorkshire breed, naturally white in color and at this stage a delicate pearl color subject dangerously to heat and to sunburn. By the time I returned, the other sows had apparently exhausted the subject of labor pains and had wandered away, possibly with the conviction that their friend was a damned fool and there was nothing to be done about it.

To rescue the pigs and get them into the shade of the barn, we used age-old tactics well known to the experienced hog farmer. Bob fetched a bushel basket and put it over the head of the sow so that she would not attack. The bushel basket blinds her and her only impulse is to back away constantly from it to free herself, and as a sow can back only slowly, a fairly agile man can keep her under control for a limited length of time. While Bob controlled the sow, Carl scooped up the little pink pigs into two pails and headed as fast as possible for the barn, but like all small pigs they set up, even at their tender age which was not more than an hour, the most ungodly racket, bringing a whole troop of sympathetic mothers to the rescue. At the same time, the mother, excited by their devilish and unwarranted shrieks of distress, leaped forward to the fray instead of backing up as she was supposed to do, demolished the bushel basket, and set out in hot pursuit of Carl. Carl is a big and long-legged fellow, but never has he made such speed as during that hundred-yard dash for the barn pursued by a whole troop of Yorkshire Niobes and Boadiceas, each one a sympathizing mother. The sounds they made were terrifying, but Carl won his race. He managed to make the inside of the barn and drop both pails and clear the barrier of a feeding creep with scarcely an inch between the seat of his Levis and the gnashing teeth of the outraged sows. Both Bob and myself had long ago cleared the fence, barbed wire and all.

At the same moment the wretched little devils spilled helter-skelter out of their baskets, stopped their shrieks and immediately began "oink-oinking!" in the most contented possible fashion while what was clearly conversation of indignation and outrage broke out among the sympathetic sows.

Such an example as this taxicab foolishness is rare indeed among sows, but this one was only young and perhaps a little giddy and like many a young woman merely stayed out a little too long.

Few animals and indeed few people are quite as shrewd or gifted with such a remarkable natural instinct as a sow … instinct with regard to what to eat and how much, how to forage off the countryside, and how to find shelter for herself and her young. Give a sow her freedom and she can take her litter right across country feeding off the land. No animal has a diet so varied and so closely resembling that of man. The pig is both herbivorous and carnivorous. Our own pigs make most of their growth off alfalfa, brome grass, and ladino clover. No animal, including man, can find a greater delight in good sweet corn, tomatoes, cantaloupe, and watermelon as many a farmer has long since discovered. Like man, pigs thoroughly enjoy meat, eggs, cheese, milk, and will even on occasion go off on hunting expeditions and bring home chickens, ducks and occasionally unwary wild animals.

But man, and in particular the mean or ignorant farmer, does not always give the hog or sow a chance to take care of their own comfort and needs. A hog is by nature the cleanest of animals and will never foul his own nest unless forced to do so by the laziness or ignorance of his owner. A sow who is either properly fed or given her complete freedom to forage off a large area has never been known to eat her young. It is only the desperate sow shut into a dark and filthy sty without sufficient protein feeds and with a lack of minerals which she cannot obtain by searching and rooting for them in the open field or swamp.[1]

The hog has followed man everywhere on his wildest pioneering ventures. Few frontiersmen ever set out into the wilderness without a sow and a boar pig and wherever he arrived, he had only to turn them loose to forage off the countryside until he could raise crops. Only an animal as big and as ferocious as a mountain lion or a bear was a match for the sow or boar, and even such formidable animals could never be sure of defeating the wits and the tusks of a boar or a ferocious old sow. Wherever man has gone, the pig has gone along to provide him with food, with litters which grew rapidly at very little cost or labor even under adverse conditions.

And no animal has ever provided anywhere near the variety of delicacies that can be obtained from the carcass of the hog. There are fresh pork and hams and bacon and sausage. There are pig's knuckles, and liver and headcheese, and tenderloin and pork chops. Indeed less of the hog is wasted than of any other animal. If I were forced to make a choice of a single meat, it would always be pork. A good steak is fine and so is a good roast beef, but for my taste both can become monotonous. Mutton is a meat that could best be described as "so-so." Lamb is a delicacy, but one can grow sick of an unvaried diet of lamb, as many a sheepman knows.

By now it must have occurred to the reader that this is one farmer who would not consider a farm a farm which did not include hogs in its program. They are the most utilitarian of animals and have justly been referred to as "mortgage raisers"—even when uneconomically and badly raised and fed.

At Malabar we are not in true corn country, although in our valley land we will challenge producers anywhere in the richest part of the Corn Belt on amounts raised per acre. We have too much hilly land and the nights are rarely hot and sometimes quite cold, even in the midst of summer. And corn is one of the most expensive crops to raise because of the high fertilizer and fitting costs and because it is a crop destructive of organic material and promotes erosion. We have always been led to believe and actually taught that corn and hogs are inevitably tied to one another.

[1] When we first took over the farms which now make up Malabar, we purchased the livestock as well. On one of the farms there was a sow shut up on concrete, about to farrow. So bad was her forced diet, owing to the ignorance of the farmer, that she was desperately tearing up the concrete with her teeth to order to obtain calcium and perhaps other minerals vital to the development of the pigs inside her.

This is a formula, a theory, and a doctrine in which I did not altogether believe, and so in our own way we began to work out a system under which we could have hogs at Malabar without corn. It worked fine and today the pigs make us more money per acre than the dairy and more than anything else save the vegetables from the truck gardens. They have also provided me personally with many hours of amusement, relaxation, and indeed wasted time, balanced at times by exasperation such as that caused by the fence-climbing pig and the big old sows who can get their snouts under a gate and lift it right off the hinges.

The program came to be based almost wholly upon alfalfa, ladino clover, and barley with no capital investment in hog housing and equipment or even in a self-feeder. It is based primarily upon trusting the shrewdness, instinct, and intelligence of the hog. In other words the system is to put everything in front of the hog that he could want so far as his dietary demands are concerned; to provide him not necessarily with the comforts of a hog's life but with the opportunity for him to create his own comforts, a trick at which he is singularly adept, and it is based upon the conviction that if the sow is given the opportunity, she knows far better than any farmer or any college professor how to feed and bring up her litters.

The grain program was based upon barley and upon winter barley, which at Malabar gives yields easily three or four times as great as spring-sown barley. We had tried growing winter barley in the early years at Malabar when the general fertility, nitrogen content, and organic material were all extremely low. The winter kill was very heavy and the yields too low to be profitable, and we abandoned the whole idea of growing winter barley, especially since we were assured by most of the neighbors that we were too far north to grow barley successfully and that the toll of winter kill would be heavy. Seven or eight years passed during which we had greatly increased the organic and nitrogen content, had built much good and deep topsoil, and immensely improved the drainage, especially in our clay soils. We took a second try and the results were astonishing. On the improved soil we found that not only could we raise winter barley, but raise it in quality and in yields so heavy that actually lodging and not winter kill became our problem.

When the neighbors said that you could not grow barley as far north as Malabar, I did not believe it for I remembered the fine crops of winter barley grown on the farm of my grandfather when I was a boy. Winter barley had been abandoned in much of the northern range of states because the soils were, on the whole, much poorer than in my grandfather's time. On many a farm the organic material was almost wholly exhausted. The texture of the soils was frequently more like cement than like good living soil. The drainage had gone with the organic material and the texture. Farmers did not plant barley early enough in the autumn and they failed to give it enough nitrogen on soils obviously seriously deficient in organic material and consequently in nitrogen. Once these conditions were corrected at Malabar there was *no* winter kill and the barley is so heavy that the fields look each year by snowfall as if they were covered by a heavy sod. Indeed, it grows so vigorously and stools out to such a degree that we have had to abandon the use of barley as a nurse crop altogether, since it simply chokes out the seedings of grass and legumes.

Barley is invaluable in our feeding program, for, although it does not give us the same weight per acre in grain as our best cornfields, it costs about a fourth as much to raise and requires less fertilizer. One has only to plant it and come back and harvest it without the cultivation or artificial weed controls which are necessary in obtaining passable yields of corn. And it is much less hard on the soils and makes a valuable contribution in organic material in the form of vast amounts of fine hairlike roots and it provides us with the quantities of straw necessary for winter bedding in our climate; straw which is converted presently into the barnyard manure that has done so much to build up Malabar Farm and each year to increase its yields and profits.

But we use barley not only for feeding hogs but for cattle and for chickens as well. It has a high content of protein as well as sugar and starch. Cattle showmen know well the virtues of steamed barley in producing a fine sheen and coat in show cattle. We are able during the winter months to achieve much the same result through putting whole grain barley into the grass silage where it ferments along with the silage and becomes brewers' or distillers' grain, the finest of all cattle feeds. Since the book *Out of the Earth* appeared, we have added molasses to the salt and barley which we habitually mixed together in the grass silage. This produces a whole and complete ration, highly palatable to the cattle, which does away with half the labor of feeding during the winter months and all the labor involved in grinding and handling grains. It is excellent alike for dry and milking cows and for young stock and if barley (or ground corn) is used in sufficient quantities in the silage, it is excellent in the production of finished beef cattle.

It is possible that but for the hogs, which I cannot do without on a farm, we might not have gone back to winter barley and re-established it as a valuable and economic feed among many of the farms in our area. Together with abundant alfalfa-ladino pasture it is the backbone of the hog program at Malabar. But the flesh and the blood of the program is respect for the hog and in particular for the sow—respect for her intelligence, her instinct, her bellicosity, and her strong-mindedness. For the whole of the program we sought merely to provide the hogs with everything they needed—to provide them with a hog's paradise, indeed a world from which they had no desire to break their way out.

The program itself is simple enough. In the winter the gilts or sows are run in the feeding barns with the beef cattle, the young stock, and the dry dairy cows. Here they are warm and dry and in one corner there is a creep where they can be fed grain and good alfalfa hay (one bale a day for every five sows or gilts). The hay which is of the best quality, closely approaching that of alfalfa leaf meal, is eaten by the gilts quite as eagerly as it is by the cattle. Each day they are given ten pounds of barley by rough measurement, or corn if you choose. Always available to them is a box of minerals including about ten minor elements, with emphasis upon cobalt, copper, and manganese which have a direct effect upon fertility and act to some extent as a preventive to brucellosis and anemia. Finely ground limestone is always available in addition to reasonable amounts of high protein supplement, usually tankage. Of this they do not consume much either winter or summer so long as the cow and steer manure is available as a source of Vitamin B12 and the animal protein factor.[2]

In addition there is at all times a certain amount of dry, strawy manure left in the feeding barns to encourage the growth of fungus and molds, which are the creators of the antibiotics or so-called "miracle drugs" that are their by-products. The hogs have a taste for these fungi and molds exactly as humans have a taste for mushrooms and morels and the molds created on or inside certain cheeses, and any close observation will

---

[2]The story of the vastly important discovery of Vitamin B12, the animal protein factor, and its relationship to growth, fertility, and resistance to disease is told in detail in the author's *Out of the Earth.* All farmers have long known that hogs which followed steers grew more rapidly, fattened out more quickly, exhibited a much higher degree of fertility, and were thoroughly resistant to disease.

demonstrate the fact that the hogs root and consume considerable quantities of these fungi produced in the strawy manure.[3]

Care is also taken, especially during the period of pregnancy, that quantities of finely ground limestone are available to the sows. The fact that the sows are allowed their freedom permits them to keep their sleeping quarters as scrupulously clean as hogs like them to be, and also permits them to have all the exercise necessary to make for a healthy and untroubled pregnancy and the healthy farrowing of litters of pigs which are strongly boned, vigorous, and healthy from the moment they are dropped.

Boars are turned in with the sows or gilts during the early winter with the purpose of bringing about the farrowing around the beginning of May. By April 15 the cattle are all out of the barns, which are left in sole possession of the sows, save for two or three steers which remain with them throughout the summer simply to make certain that the sows have always an ample supply of Vitamin $B_{12}$.[4]

During the winter months the sows are also permitted to roam at will in the lots which are used for rotated pasture during the summer months. This again permits them really unlimited exercise and the freedom so necessary to the happy, healthy hog.

Owing to the balanced, free-choice diet of the sows, they do little if any rooting save in the dampest spots of the pasture lots where they find earthworms during the hot summer months. In fact, they do so little

[3]Nowadays it has become the fashion to put various antibiotics into the prepared feeds for hogs, a concession to the fact that hogs actually like and even crave these things and that if given the opportunity, they will find them on their own. The antibiotics are put into the feeds to promote rapid growth and to some extent health, but we are still in ignorance of the *exact* amounts which are necessary and harmless, and it is quite probable that these amounts vary according to the metabolism of the individual hog. In many cases the use of antibiotics in prepared hog feeds has produced lame and crippled hogs because growth has been accelerated beyond the point of *healthy* growth and produced soft and *unhealthy* bone structure. When antibiotics are *mixed* with hog feeds, the hog has no choice as to the amount he consumes and is not permitted to employ the very strong instincts which make the hog under proper conditions a strong, healthy, and self-reliant animal, able to take care of himself and arrange and balance his own diet better than any farmer or professor can do.

[4]Cattle are the greatest producers of Vitamin $B_{12}$. in the world. It is manufactured in their stomachs, provided there is no deficiency of cobalt necessary to the molecular structure of the vitamin. If there is a deficiency of cobalt, the animals will not breed properly, grow slowly, and be subject to certain diseases, especially those affecting the reproductive glands and organs. If the deficiency is great enough, they will develop acute anemia, refuse to eat, and simply die. The details of this relationship are explained at length in the author's earlier book *Out of the Earth*.

damage that the necessity for "ringing" them to prevent rooting has never arisen. No hog is ever "ringed" at Malabar. In fact, what little rooting they do in the alfalfa-ladino pastures serves to cultivate and stimulate the growth of the legumes during the following summer. In the early spring, all the rotated pasture lots are worked over and leveled with a field cultivator, fertilized, and treated with new light applications of alfalfa and ladino seed, so that they are always in maximum production of forage during the summer months.

In the barns once the cattle are turned into the fields, fresh straw is added to the accumulated steer manure and in this the sows make their nests for farrowing at will where they like. At this time they are given no special attention whatever; they are given no slops, mashes, or special feeds. They are simply provided with everything they need and left to themselves.

The loss of newborn pigs from overlying is negligible, indeed only a fraction of one percent, probably no higher than that in many an elaborate and fanciful farrowing house put up at great expense. After two or three days the sow simply conducts her pigs out into the lush pasture of the rotated fields and from then on the whole family is on its own. From time to time, a quantity of barley is thrown onto the ground and is replenished when the sows and pigs have consumed it. Save for replacing minerals and limestone once or twice during the summer, that is the only labor involved. When the sows have consumed the young, nonfibrous ladino and alfalfa, the gate of that plot is closed and the gate of the adjoining one opened and the first plot is clipped short to promote a fresh tender growth. This process is repeated over three rotated plots several times a summer.

It is notable that the sows and pigs pasture these lots quite as enthusiastically as cattle would pasture them and that the milk production of the sows is notably very high. The pigs are never weaned but allowed to run on the sows until they no longer give any milk.

Under this program we have no capital investment whatever in housing or feeding. We are able to put pigs on the market weighing around 225 to 235 pounds in from six and a half to seven months. They are lean, fastgrowing pigs and being of the Yorkshire breed or purebred Yorkshire crossed with Hampshires, bring top market price. On such a program we are able to make more profit out of one litter of pigs than most farmers with large capital investments in hog houses, heating equipment, concrete, and heavy labor costs make out of two.

It should be remarked that these hogs live in a kind of hogs' paradise. They have during the hot summer months the shade and cool of the feeding barns, and in the winter months shelter from cold winds and the natural heat produced by the cattle and the radiant heating which is constantly being generated by the heat of the accumulated manure beneath the fresh straw bedding. At all times they have fresh, cold spring water which supplies all the rotation lots and they have clean mud wallows through which the spring water runs. When a sow reaches a weight of five hundred pounds or more she is disposed of, as her great size makes her unwieldy and likely to overlay some of her pigs. This is not the sow's fault. Nature never meant sows to weigh up to seven or eight hundred pounds. That was man's doing, which he accomplished by selective breeding.[5]

Our gilts, being Yorks which tend to have good-sized litters, average between eight and nine healthy pigs. Perhaps the most noteworthy fact in the whole of this experiment in which everything possible is left to the pig himself, is that in the five years of the experiment we have never had a sick pig. Nor has any pig ever been inoculated or vaccinated. As I have observed elsewhere in this book, we are not cranks at Malabar and have no objection to using dusts or sprays, vaccinations or inoculations *when necessary*, but we expect to carry along as far as possible *without* them, feeling our way toward the efficacy of balanced soils, proper nutrition, and proper care to build up resistance naturally in both plants and animals.

I have observed that, among my friends who raise hogs, the ones who have the greatest troubles from sickness, sterility, rheumatism, pneumonia, and contagious infections are likely to be those who do not watch the balanced diet or living conditions of their hogs or who have the very greatest expense and investment in daylight lamps, heaters, elaborate farrowing houses, disinfectants, etc., etc., and again etc.

Among all farm animals the hog resents a curtailment of its freedom more than any other and, as every farmer knows, is the most difficult to

---

[5]In the early stages of this experimentation we raised thirty-one pigs from single litters of each of two enormous sows. One had seventeen and the other fourteen living pigs and neither overlay one of them. We took five pigs away from the sow with seventeen and two from the one with fourteen. They were given to three of the children on the farm who brought them up on a series of bottles and nipples arranged in a row in a wooden rack. Needless to say, they became first-class pests who set up an ungodly squealing every time one of the kids appeared.

keep confined in any small area. The hog also has an immensely strong instinct and intelligence with regard to taking care of himself and his own needs. Shut him up and no amount of trick diets, concrete, or sunlamps can make up for his lack of freedom and his right to forage for himself. Not only do these elaborate and expensive methods of hog production on some farms fail to prevent disease, poor litters, and a lack of vitality, I am not at all sure that the hog, an immensely intelligent animal with a great liking to roam, does not suffer psychologically from being closely confined and that the depression arising from his close confinement does not have some effect upon his general health. Unlike the stupid sheep which will lie down and die under adverse circumstances with no will whatever to live, the hog has an immense will to live and to keep his freedom.

During 1951, I acted as moderator on a series of thirteen panel programs investigating the costs of foods conducted over a major radio network under the auspices of *Successful Farming* magazine. During the panel concerned with the cost of hog raising and of pork, one Iowa hog farmer produced irrefutable figures that showed his profits upon hogs farrowed and raised on waste swampland on his farm were considerably greater than most corn-hog farmers following the conventional methods. They were startlingly greater than the corn-hog farmers raising two litters a year with heavy capital investments in elaborate hog installations and heavy labor costs that went with them. Also the health record of his hogs was better.

In 1955 we are embarking at Malabar on a greatly extended hog program which involves the enclosure of about 150 acres of swamp and forest land now wholly unproductive save for an occasional cutting of timber. It is our plan to permit the sows and their litters complete freedom within this area, which is also furnished with water and clean wallows by a large spring creek which maintains an even flow under all conditions. The hogs will also have access to rotated pastures of alfalfa and ladino, with a single feeding lot of two acres where they will be fed whatever grain they need, and where the pigs can be rounded up and loaded at marketing or sale time. Within the wooded and swampy area they will have access to an even greater variety of diet and freedom than under the present system. As the gravel in the spring stream is of glacial origin and about 40 percent partially broken down dolomitic limestone, even the question of minerals tends to solve itself. We are planning to run in the area a hundred or more sows for a beginning and will sell the pigs off at weaning time, re-breed the sows to be sold in the autumn to other farmers, close down all operations and eliminate expensive grain feeding until January first, when a new lot of gilts and sows will be purchased and put into the feeding barns with boars to breed for farrowing in early May, when the same routine will be followed in the succeeding year.

The new plan (that of selling the pigs as weanlings at two to two and a half months and re-breeding the sows for autumn sale) promises to be even more profitable. During the past four years the weanling pigs we have sold never brought less than $16.50 and some brought as high as $20.00. These weanlings have consumed only very small quantities of grain and have been raised almost wholly on milk and legumes. As our average litters run between nine and ten living pigs, the return per sow is very high, especially when the sows are sold as bred sows in the autumn. Bred purebred sows or registered gilts in our area are, at the time of writing, bringing a hundred dollars apiece and ordinary grade sows are bringing eighty dollars upward. These figures are important, especially in view of the very low capital investment involved, the virtual nonexistence of veterinarian bills, the elimination of the expensive mashes and supplements, and a labor cost which is extremely low ... merely a part-time job for one man. I mention these economic factors as they are of great interest to the practical farmer.

In all the experiments in raising hogs at Malabar we have been concerned primarily with health, simplicity, and *net* profits to the average farmer. It does not matter that a farmer puts ten thousand hogs on the market if he makes little or no profit and if he is plagued by sickness and ills of all sorts. A farmer raising a hundred hogs with common sense may have a much larger net profit. The truth is that the whole technique of raising hogs as advocated by many "authorities" has become so complex and so complicated that both the margin of health and the margin of net profit has been greatly lowered for any farmer who attempts to follow out all their advice and instructions. Capital investment in elaborate housing and feeding installations make it necessary to raise hogs for years before the investment can be amortized. The costs of daylight lamps, of artificial heating, of veterinarian services, of inoculations and vaccinations, of expensive hog "mashes" and artificial chemical "stimulants", and a hundred other things have made of the comparatively simple operation of breeding, raising, and feeding the most intelligent and self-sufficient of animals something which, if carried out in all the details, not only devours the profits but greatly increases the labor and in general produces a kind of nightmare not only for the hog but for the farmer who attempts to follow all the advice that is put out. Much of this comes frequently through high-pressure advertising campaigns put on by chemical, farm equipment, and feed companies and frequently by "experts" who have never possibly imagined themselves to be hogs or what a hog would like in order to be healthy, comfortable, and happy, nor have they ever at the end of the year been forced to balance costs against profits. They are men who regard hogs as something to be kept in a test tube rather than as highly intelligent and self-sufficient animals. A little time "wasted" leaning on the fence of the hog lot where the hog is happily taking care of himself might be of profit.

In much of the farm picture the nation has gone more and more toward flouting natural law and espousing short cuts, artificial tricks, and in general a whole panoply of artificialities which have simply brought upon us new and greater costs, new dangers of disease, and sometimes even new diseases.

The cost of feeds is one of the most important factors in the raising of hogs and in coming back to this, we come back again to the primary factor in all farm economics. How much do you raise per acre and what is the analyzed quality of what you raise? The farmer who farms five acres of corn or barley to produce what he should be producing on one acre will never make any money in hogs. The more hogs he has, the more money he will lose. Only a well-kept set of books will prove that to him, but the farmer who farms five acres of land to produce what he should on one, rarely, if ever, keeps books. That is why heavy farm supports are politically necessary.

From all of this the reader has perhaps gathered the belief that the writer has a higher opinion of some hogs than he has of some people, and in this I am compelled to say he would be quite right.[6]

---

[6]There is an excellent and fascinating book called *Pigs From Cave To Corn Belt*, dealing with the contributions to civilization and the opening up of new country made by the porker family. The authors are Charles Wayland Towne and Edward N. Wentworth. It is published by the University of Oklahoma Press, Norman, Oklahoma.

## IT CAN'T BE DONE

"There aint no such word in the dictionary as can't."
—*Old Saying*

O NE OF THE MOST INTERESTING ASPECTS OF AGRICULTURE IN OUR TIME
is the number and variety of things which "can't be done." It is a cry
that I have heard countless times concerning feeding programs, the
restoration of soils, the management of pastures, the diminishing use of
dusts and sprays of all kinds, the prevention of cannibalism in chickens,
the special uses of certain farm machinery, new types of tillage, and so on
*ad infinitum*. It is a cry which comes from feed manufacturers, from the
advertising and public relations agents of big chemical companies, and
not infrequently from the campuses of our colleges of agriculture.
Usually it is born of a lack of imagination.

At the same time that the cry "It can't be done" is raised, many of the
same agencies and companies which raise it harass the farmer with salesmen, pamphlets, bulletins, and propaganda of all sorts advocating
panaceas and shortcuts which the sensible and practical farmer knows
well enough cannot succeed and which will appeal only to the farmer
who is perpetually looking for such things—the farmer who is seeking
the quick, easy way in the same fashion as the speculator who is always
looking for the quick, easy buck, and never finds it.

For the past two or three years the whole network of roadsides in the
vicinity of Malabar Farm has been disfigured and made hideous by the

use of chemical sprays which were *supposed* to kill off *all* roadside vegetation. The salesmen and the advertising men of the big chemical companies had sold the county commissioners a bill of goods. In almost no
case was the vegetation entirely killed. The sprays only maimed and ren-
dered hideous the roadside and made the whole of one of the most beautiful of countrysides appear simply as if it had been hit by a terrible
blight. Even lanes and back roads were disfigured for the whole of the
summer, and even before the end of the summer the vegetation began
growing back again. The next year the process had to be done all over
again with the same evil results. For the whole of the season, the hideous-
ness was far worse than that created by any obscene roadside sign. And
there were other evil effects in the killing of vegetation inside the adjoining fields and the utter destruction for a time of the game cover so valuable in keeping up the supply of game for hunters. In the autumn, instead
of the brilliant foliage that borders our roadside, there were only the ugly,
dead branches of the half-slaughtered trees and shrubs. And no one has
yet been able to predict definitely what the effect of such spraying year
after year would have beyond that of creating in the end an utter desert.

A few farms, including our own, had the courage to protest, and in the
following season every farm was given the option of having the adjoining
roadsides sprayed or not sprayed. The response was interesting. The good
farmers objected; the indifferent or bad farmers did not care. They still
believed in naked wire fences and no vegetation but cash crops. In many
a field, the roadside growth would have mercifully hidden the thinness of
their soils and the poverty of their crops.

This year (1954) the spraying was abandoned in the middle of the season. I do not know whether this was because of its inefficacy, or because
of widespread protests from those good citizens who seek to guard the
decency and beauty of our countryside. Nineteen fifty-four was an elec-
tion year and it may have been that the county commissioners found it
good politics to employ the extra men needed to cut down the vegetation
by hand.

The effects of the indiscriminate use of even such sprays as DDT are by
no means as yet properly or accurately observed and assessed. Many a
farmer whose poor and unbalanced soils produce sickly crops which
actually *attract* the whole beetle family, uses DDT indiscriminately in an
effort to correct the results of his own fundamental failure. The effects

upon insects, birds, and fish and the upsetting of the whole natural system of checks and balances is of no importance to such farmers, yet when the reckoning comes they are and will be the loudest in their wails for help.

As has been pointed out elsewhere in this book, Missouri and Kansas State agricultural colleges have discovered that an extra application of nitrogen is more effective in protecting corn from chinch bugs and wheat from green bug than any amount of spraying with DDT. Also it will greatly increase the crop yields for the farmer with nitrogen-poor soils. It is probably true that 90 percent of all crops raised in the U.S. are constantly starved for nitrogen. But the effort of controlling attack by certain specific insects has thus far been an effort to suppress the attacks after they occur by what might be called "artificial" means rather than to prevent it by fundamental practices evident in all sound and really modern agriculture. Like much of the trend in human medicine, the greatest part of the effort is directed not at producing healthy and resistant plants or in the case of veterinary and human medicine healthy and resistant animals and people, but at patching up plants, animals, and people *after* they become sick or subject to attack.

At Malabar Farm we have never been *doctrinaire* nor have we been cranks. We are primarily interested in what works and what is permanently good for the farmer, his soils, his crops, and his animals and what gives him the highest production and the best nutritional quality. We are constantly experimenting and feeling our way, and in doing so have arrived at some very interesting results. We do not always pretend to know why; it is frequently the job of the laboratory research man to discover the reasons. Nevertheless we have our theories and these have led in a single and apparently inevitable direction—that nature has provided the means of producing healthy and resistant plants, animals, and people and that if these means and patterns can be discovered and put into use, the need for "artificial" and curative, as opposed to preventive methods, is greatly reduced. In any case we are convinced that in a sound and permanent agriculture there is no such thing as a panacea or shortcut. Indeed the whole of agricultural and livestock science and even human medicine, if sound, is merely the business of discovering certain natural patterns already in existence, putting together the various pieces, and discovering their relationship to the whole universe; indeed, such a process is science itself.

All the true and lasting and sound contributions to agriculture, horticulture and livestock breeding have inevitably come within this category. The discovery of the antibiotics and their uses, as we know them, is no more than the discovery and understanding of one of nature's soundest patterns ... the universal operation of the fundamental law by which all of us, and the animals, birds, and fish and vegetation about us, are still on this earth and capable of reproduction ... the law of birth, growth, death, decay, and rebirth. So indeed are the fungi and molds from which the antibiotics are derived and which bring about the decay and dissolution which restores everything again to the earth from which everything is again reborn and renewed. This process and pattern includes too the beetles and all the scavenging insects whose place in the universal schematic law is the reduction and destruction of the dead and dying material, which without them—the fungi, the molds, many species of insects, the bacteria—would today clutter the whole of the earth to the depth of many feet.

If there had been more research and more "specialists" occupied in research into this vast, complicated, and fundamental pattern and fewer concerned with finding and creating every sort of new poison and chemical shortcut *after* plants, animals, and people become sickly and unproductive, agriculture, horticulture, and livestock enterprises would be much further along upon a truly sound basis of health, quality, blood lines, production, and economics.

Perhaps the most ludicrous and obvious of the efforts to suppress symptoms rather than to prevent sickness, has been the advice given out by poultry "experts" who sought to suppress cannibalism in chickens. Virtually all of the research and much of the advice to the long-suffering poultrymen have been along the line of finding the means of *frustrating* and *suppressing* the morbid cravings of the chickens rather than to find the reasons for that morbid craving and correct them. These same "experts" have advocated painting the windows of the poultry house red so that the feathers of white chickens will appear to be red like the flesh of the hen which has been attacked by cannibalistic sisters. They have advocated putting false beaks and even red glass spectacles on every hen, and lately, even *after* the discovery that cannibalism was caused by a deficiency of the animal protein factor, or Vitamin $B_{12}$, in the diet of the hen, the "expert" at one of our great agricultural colleges advocates the cruel practice of clipping the beaks of every hen so that she will be unable to practice cannibalism upon her sisters.

I repeat that in nearly all of this research into cannibalism, the effort was consistently directed toward suppression and frustration rather than toward finding the basic *cause* and correcting it. Indeed, much of the research is on the level of that done by the alchemists of the Middle Ages.

This same approach has long been in evidence in many of the troubles connected with the livestock industry and it has extended even into soils and the research concerned with them. The meteoric rise and fall of the propaganda and advertising and general exploitation of "soil conditioners" is evidence of the quick failure of another shortcut that was intended to by pass natural laws and produce a quick buck. Apparently it never occurred to any of those concerned with the soil "conditioners" that if soils were in desperate need of "reconditioning" in order to get proper germination and permit the seedlings to emerge and to make sound root growth, these same soils would be in such poor condition that normal, vigorous plant growth *after* germination would be impossible. The average technical chemist, the average advertising man, or chemical salesman could not be expected to know or properly understand all this, but the agronomists hired by the chemical companies at high salaries should have and possibly did know it and should have said so. It is possible that the high salaries paid by the great industrial corporations—many times higher than any paid by universities or colleges—have tempted many a college-bred agricultural technician simply to become a "yes man" to any idea, however fantastic, that may come from the top office of the advertising agency.

In the case of the soil "conditioners" the economic aspect in relation to practical agriculture seems to have been overlooked altogether; the use of these dubious panaceas would have cost the farmer many hundred dollars an acre to secure flash results where a few crops of green manure or a little barnyard manure would have produced *permanent* results at a cost of a fraction of that of the soil "conditioners." For two or three years we heard about the miracle work of soil conditioners everywhere, in newspapers and agricultural magazines, and then suddenly a deathly silence. But the big chemical manufacturers find themselves constantly in possession of new byproducts, some of them of a violently poisonous nature, for which they wish to find a market. Unfortunately the farmer and gardener and livestock man are all too frequently the victims of a campaign to palm off these materials as shortcuts and panaceas. Frequently very little science or research is involved in the process, but a great deal of crude commercialism and exploitation.

Perhaps the most annoying factor is that many of these paid experts and many of those who seek to suppress symptoms rather than to prevent the disease or ill, cling frantically to the title of "scientist", however superficial or false may be the basis of their efforts. Science and research to them is very frequently only the process of finding means and nostrums for the suppression of symptoms. Yet these same "scientists" seek to discredit frequently the methods and results of those explorers into the realm of the sound and natural patterns set up by nature herself and into the realm of the very laws of the universe. Perhaps it is more profitable to operate in this way, but it is not better science.

It is true that many a sound worker in the realms of agricultural, biological, and nutritional research works under great handicaps. As a friend of mine on the staff of one of the really excellent experiment stations in the country once said, "You don't know how fortunate you all are at Malabar. If I want to make any important experiment I must first get permission of the dean of the college. Then I must secure an appropriation. After that, if I fail, the failure goes down as a black mark against my name and reputation. And in addition there are not a few fellows around me who are hoping that I will fail. At Malabar you do exactly as you please and finance your own experiments and research, and if you fail to get the results for which you hoped, you merely say so and forget it and go on to something else. You don't even have any campus politics."

But perhaps the most extraordinary of attitudes is that of the "expert" who feels that everything in the world ... or that small part of the world with which he is concerned ... has already been discovered, and that there are no changes, modifications, or advances which are needed or can be made. This is indeed an extraordinary attitude in the face of the infinite complexity of the universe and the times in which we live when new, revealing, and revolutionary discoveries of patterns or parts of patterns in the cosmic laws are being discovered so rapidly that it is virtually impossible for them to be recorded in any ordered, coordinated, and related fashion. Yet I have listened, not without wanting to get up and shout, to a long discourse by a "specialist" with a certain following and prestige to prove that everything has been discovered which could possibly be discovered about soil bacteriology!!!

The greatest handicap in all our agricultural research and experimenting today is the *lack* of imagination and of the passion for exploration in the vast areas of knowledge and scientific fact which are as yet virtually untouched. In some of our agricultural schools textbooks are actually in use which teach methods that have not only been proven unsound through the evidence of newer discoveries, but which have actually been proven damaging and destructive. But it is a slow process to change textbooks, and perhaps the greatest handicap is that it is also expensive; so many a young man today learns in college a lot of things which not only bear little relation to modern and truly scientific agriculture, biology, pathology, economy, and a dozen other fields, but which are simply not true or scientifically sound.

The overspecialization which is the greatest fault of American education in almost every field is by no means absent from agricultural education. It has tended more and more to separate agricultural research and experimentation from the earth itself and to confine it to the test tubes of the chemical laboratory. Broadly and fundamentally speaking, the earth itself is the source of all agricultural knowledge and the research laboratory exists to discover *why* certain processes take place and *why* certain laws operate within the earth and consequently within the universe. One does not *begin* in the laboratory to discover natural laws governing soils and patterns of growth, resistance, health, and in general the overall operation of natural law. It begins with observation, experimentation, and imagination operating in the field actually existing *within* the earth. The shortcuts, the panaceas nearly all come out of the test tube of the chemical laboratory; and certainly half the sound contributions and advances in agriculture, horticulture, and the livestock industry have come *not* out of the laboratory but from the smart farmer and livestock man who lives *with* his soils and his animals, observes them and employs his imagination and powers of deduction. Later on the laboratory may find the answer as to why a given practice produces notable results, but the original spark must be ignited in the field and the idea must come literally out of the earth.[1]

[1] As the experimenting and research in soils at Malabar progressed and increased, it became necessary to add experts to the staff, experts who were not trained to behave as if the laboratory and the soil were two separate worlds, but men who were trained in *all* the techniques and at

Over the years we have come at Malabar to find an almost malicious pleasure in doing what "can't be done." Sometimes we have done it by accident, sometimes by design based upon observation, knowledge, and imagination, but we have done it many times and in many fields. At Malabar there is Mount Jeez, which was known for two generations as "Poverty Knob" and would not on its hundred acres produce enough pasture in a year to feed a half-dozen cattle for a week. When we took it over, the hill was a desert of broom sedge, poverty grass, blackberry bushes, and every sort of low-grade weed. For a long time now it has been producing good pasture which in a season of good rainfall will carry a head to an acre and a half or two acres. Our best, fastest-growing young stock pass their summers there, without any grain whatever.

We had to find out not only how this land could be restored and put to quality production, but how it could be done quickly and cheaply. The working out of such a process required knowledge, observation, outside imagination, the proper kind of machinery, and the proper legumes and grasses. After much hard work, and some money, we did evolve a system which met these requirements. If there are doubters, we merely repeat our invitation to come and see, count the acres and the cattle and observe their condition, even in a drought year. In evolving the process, which was later applied generally to all the poorer lands at Malabar, we have also built many inches of good topsoil where before there was none.[2]

---

the same time were not opposed to taking off their white jackets occasionally, putting their hands in the soil and observing the plant *not* in the greenhouse or the laboratory but in the field under *natural* conditions. I could find in this country no expert trained to see the whole rather than one or two of the parts save a few already engaged in good jobs, usually in a family commercial vegetable or fruit business and who were consequently unobtainable. In this group some had been to college, some had not, but they treated the soil *as their laboratory*. In consequence I was forced to go to England where the kind of training and experience I was seeking is a tradition and a practice. I obtained two young men, one graduated from the North Wales Horticultural College and the other trained in famous Kew Gardens in London. Like myself they look upon actual work in the field and *with* the soil not only as honorable and even distinguished, but as absolutely essential in any real training in agriculture and horticulture and in getting sound results. They have added immeasurably to the research, efficiency, production, and general welfare of Malabar Farm and are primarily responsible for the entrance of the farm in the vegetable and flower business. They look upon everything they produce as an instrument of agricultural and horticultural research and their trained powers of observation I have seldom seen surpassed. They *live with* the soil and the plants which grow in it. This kind of thing is extremely rare in the U.S., where each little facet of the whole vast field is specialized and narrowed sometimes to the point of bigotry. There is little respect for that all-encompassing knowledge based upon the fact that no one element of the universe is isolated and unrelated to all other elements.

The Soil Conservation Service men say that it took nature ten thousand years to build an inch of topsoil and they are quite right, but it does not take man that long and today at Malabar where only a few years ago there was only bare eroded topsoil, we have topsoil three to seven inches deeper than the original virgin topsoils and far more productive than the original soils. It was accomplished by the use of grasses and legumes imported originally from other parts of the world and some of which grow more vigorously here in our Ohio country than in the countries from which they came. It was done by the use of *deep-rooted* grasses and legumes, by the intelligent use of lime and chemical fertilizers (translated eventually into the organic form of green manures), and by the use of modern power and deep-tillage implements now available to any farmer. On many wasted, once abandoned acres, we are actually today growing the finest of truck garden products ... celery, tomatoes, lettuce, eggplant, cantaloupe, what you will. Most of our fields, even on the steeply rolling land, have the structure, quality, and fertility of truck garden soils.

And the reader must bear in mind that none of these results were obtained by the expenditure of great sums of money. There has always been one ironclad restriction at Malabar—that nothing should ever be done there which *any* farmer could not afford to do. The thousands of farmers who visit us each year know that this restriction has been kept. You cannot fool a wise and practical farmer. Malabar has never been a "show place" either in the buildings or in the expensive trappings of the millionaire's farm in the old days, but it is a beautiful place with its hills and valleys and forests and springs and streams, where we are willing to challenge the production per acre on any crop anywhere in the U.S.

The other slogan has always been "Come and see for yourself." Nothing is hidden at Malabar. During the week, visitors are simply turned loose to go anywhere they like and ask any questions they like from any man working on the place. On Sunday afternoons there are tours from May 1st to November 1st, conducted by the proprietor or by one of our

---

[2]In the process we worked closely with the Sunnyhill Coal Company, one of the large strip mining companies operating in Ohio and Pennsylvania and its president, Cliff Snyder, who had a great and intelligent interest in finding practical methods for the restoration of spoiled bank lands and some thousands of acres of so-called "worn-out" farmlands in possession of the company. The methods used at Malabar have since been applied for years with success and have spread among other strip mine operators throughout the Middle West. The details of the method are recorded in detail in *Out of the Earth* and touched upon elsewhere in this book. The contacts with Mr. Snyder also led to the work and experimenting done later at Malabar with the use of minor elements and of soluble fertilizers through irrigation.

Malabar staff, where many of the things done there are explained in detail and any and every question is given a direct and honest answer. Each year, more than twenty thousand visitors, mostly practical farmers, as well as scientists and research men and women, pass through Malabar. On the occasion of the Field Day to demonstrate the rapid building of topsoils, which was sponsored by *Successful Farming* magazine, more than eight thousand farmers with their families, coming from twenty-seven states, visited Malabar. It was an experience I should not like to go through again.[3]

If at times there appears within these pages what appear to be traces of acrimony or irritation, it arises from the fact that Malabar has from time to time been subject to somewhat strange criticisms from some of the more hard-shell elements in academic circles. One such criticism wag repeated to me not long ago in the following words: "It would be fine at Malabar if they only told the truth!" This came oddly enough from a professor at a state agricultural college who, so far as we know, had never set foot upon our land or even seen the place from a distance. I also doubt very much whether the gentleman in question had ever had his hand in the soil during the past thirty years.[4]

Perhaps the oddest of reactions is that of the academic mind, which standing with its feet in our soils in the midst of a richly producing field seeks every sort of reason for the success save the reasons which are obvious and which we know are the true ones, frequently reasons which we have discovered ourselves on our own or by putting already available knowledge into new and slightly different patterns. The implication is

[3]The *Successful Farming* day will long be remembered in the area, for the backcountry roads were filled with traffic, the cars running bumper to bumper. The neighboring village of Lucas suffered an all-day traffic crisis beyond the powers of the village marshal to control. The demonstration was scheduled for 1:30 P.M. but there were hundreds of cars already parked in the rolling meadows by eight in the morning. Members of the staff of *Successful Farming* provided 2,500 programs which were all gone by 10:00 A.M. They endeavored to check and keep a record of the cars, an effort they were eventually forced to abandon. As the visitors were not expected until the afternoon, no provision was made with the local churches or granges to provide food, so the visitors were forced to forage for themselves. By evening the stores in the surrounding villages were depleted of everything from breakfast foods and tinned soups to cigarettes. The demonstration took place on a Saturday afternoon but many visitors were still there on Sunday and the following Monday.

[4]On two or three occasions we have discovered what might be called espionage expeditions by more academic-minded individuals when they were found wandering about the place, without our knowledge of their presence, poking here and there. In one case we found two men with an elaborate soil-testing apparatus operating on the outskirts of the farm. One of these trespassers later became one of our best friends and strongest cooperators.

that we do not know what we are doing and that much that has happened has been an accident or could not work out on a different kind of soil. The simplest answer is that we at Malabar, along with countless other intelligent farmers throughout the U.S., have frequently been doing what some teachers and "scientists" should have been doing for some time past. Frequently we have been doing things which "cannot be done."

It is true that frequently some of our critics have been handicapped by restrictions from which we do not suffer at Malabar and from which no independent farmer suffers; but the fact remains that all the restrictions, all the criticisms, all the jealousies of the academic world did not hamper or frustrate Pasteur or Fabre or in our times Ed Babcock or Firmin Bear or William Albrecht or George Washington Carver. These were all men who did what "cannot be done" and proved what "could not be proved." The tragedy is that again and again we find the highest positions of authority in many of our schools and universities are occupied by men of rigid mind, without imagination or the fundamental vitality which lies behind all true scientific advance and research, men who are lacking not only in imagination but even in curiosity, who are without discontent and are happy with things as they are, since imagination, speculation, and curiosity are uncomfortable and even dangerous elements which upset the even tenor of lives in which, by being safe, a salary will go on forever with a pension at the end. This is essentially the mentality of the hack government bureaucrat everywhere and there is far too much of it in our agricultural education today.

But behind the unimaginative, congealed mind there lie other more subtle factors of tradition, of education, and experience. The great scientist, the great painter, the great creative artist is always an essentially simple man with the capacity to advance, unhindered by conventions or limitations of any kind, straight to the point, because it is the point, the end result, which is his principal and indeed his only concern. He is not especially concerned merely with the tag of an academic degree (although he may on the way acquire great quantities of these). The academic degree does not necessarily make him a great thinker or a great teacher; it means merely that he is able by determination and persistence to acquire and retain, beyond the point of examination or the writing of a thesis, certain facts and figures. It does not make of him a great teacher or a great research scientist; for these things other qualities of a more fiery nature are required in addition to the acquisition of knowledge. Yet the acquisition of these tags has become increasingly not only the primary requirement, but the primary consideration of American education in nearly all fields. With the result that in many of its aspects American education has become hidebound, limited, overspecialized, and in its more extreme manifestations, a kind of living death.

The great scientist and the great creator, indeed the great man, is not primarily concerned with academic honors or campus politics or a salary to be counted upon, which is followed by retirement pensions or by keeping up with the Joneses or any of the things which confine and plague the ordinary limited mind. And, of course, the tragedy of this limited mind is that it is wholly unaware of the difference between fire and deadwood or between achievement and mere academic recognition. His very goal is frequently *not* the achievement but the academic recognition; and so he is perpetually blinded and limited, while at the same time he is uneasily aware of his error and suspects that far beyond and above this lies something which he does not quite understand and which he will never quite achieve.

It is difficult sometimes to establish in the minds of the mediocre the difference between real simplicity, which is born of intelligence, fire, vitality, and the elimination of the nonessentials, and what might be called the simplicity of the "hick", who is merely simple because he does not know anything outside the immediate small area in which he lives and functions. With this "hick" quality, which implies a lack of worldliness and of a balanced sense of values and of experience, our educational scene is richly endowed, not only in its agricultural, but in all its aspects. Some of the men I have known, with all sorts of degrees, have achieved them by sheer persistence and plugging hard work, and yet despite them still possess this peculiar "hick" quality which is the essential mark of the limited and mediocre. Perhaps I am being oversubtle; what I am trying to say is that over the years I have acquired a kind of dread and horror of much that is associated with campus life, with its limitations, its politics, its jealousies, its "life in death" self-satisfaction, and the aggressive smugness born of a sense of inferiority. Perhaps this is because I have been singularly fortunate in having been able all my life to lead the independent life of an individualist, but in the end such an advantage may not be so much good fortune as determination and free will, and the will not to seek refuge in conformity and safety.

Let me add again at this point that despite a certain savagery in the above paragraph, the writer is really a very amiable fellow, very fond of people, and with countless friends in the world of agriculture and agricultural education.

I think very few city or town dwellers have ever appreciated the actual heroism of the average good farmer in carrying on and even increasing production of food for ourselves and for our allies during the interminable period of the last war. One had to live through the period as a farmer to understand the long hours, the endless worries, the bitter hard work of both farmer and family, the slipping downward of efficiency despite superhuman physical effort; something which can be infinitely depressing to the point of inducing permanent melancholia. Nor can the city dweller appreciate or understand the many things which "can't be done" which the farmer accomplished during the harrowing time of war.

No city dweller can understand the misery of having to accept what could only be described as rejected "white trash workers;" rejected elsewhere not because of any physical disability but because they preferred to quit work altogether as soon as they had fifteen dollars in cash, because they were quarrelsome not only with their employers but with their own families, whom frequently they threatened with guns and knives. The capacity of such help for producing crises and bedlam on any farm is, as any farmer could tell you, endless. Any farmer knows that even with good help and efficient operation, there is practically a new crisis every morning on a farm. With "war help" there were many crises each day. The good farmer went to bed each night troubled and worried and awakened each morning to find his troubles and worries had become realities overnight. There were certainly moments in the lives of many a farmer when he would gladly have changed his place for one in the front lines of the trenches. Certainly they worked much harder and suffered more than the great majority of generals and admirals and bureaucrats. And when the Korean mess finally came to an end, the farmer found that his prices kept steadily going downward while the prices of everything he had to buy had increased as much in some cases as 200 percent. And still, even in time of peace, the armed forces through the so-called "Selective" Service continued to conscript his sons and harass his operations, as it still does at the time of writing.

Of course, the greatest and most damaging effects of all these difficulties at Malabar was that it set back *all* operations on a farm which was engaged not only in the difficult business of farming but in the work of experimentation, research, and restoration as well. In addition to farming, we were always engaged in two or three other jobs as well, and during the war years the experiments in soil in relation to the nutrition of plants, animals, and people limped along, much of it carried out by the muscles and strong back of the proprietor himself. In all of these fields and in the general rehabilitation of soil and production, Malabar, like countless other farms and many research agencies, was set back in its progress by several years. And so the general appearance of the farm suffered.

Such a statement is not made in any sense as an excuse or even as an apology. The fields of Malabar, growing each year more productive, looked better than the majority of American farms, but there was a constant sense of dissatisfaction, especially for the perfectionist, who knows that perfection is never wholly achieved but that the striving toward it is of the utmost importance. Many a reader and especially the good farmer will understand the steady, unrelenting sense of depression and even desperation which arises from the sight of the jobs that need to be done and cannot be done, of the tasks which are carried out sloppily and without a sense of responsibility, but casually and lazily, merely in order to earn enough money to eat. Only the toughest or the most dedicated farmer or livestock man can take this sort of depression and discouragement indefinitely. During and after the war, many did not: they abandoned agriculture for work that was easier, more banal, and relatively free from worries and responsibilities.

At this point it is impossible in a book written with intentional honesty not to comment upon one sad fact which has contributed frequently to failure, to mistakes, and to ragged-looking fields from time to time at Malabar. This is the poor quality, the inadequacy, and the lack of responsibility which is manifest in so much of American farm labor, even among what might be called skilled labor and college-educated young Americans. We have had at Malabar a long experience and opportunity for observation and comparison with the workers and young farmers of other nationalities, for we have had from time to time Danes, British, Swedes, French, and Brazilians working with us for periods of months and occasionally years, and we have had countless other native-born American boys and young men during the summer harvest season. In addition to these, I have had experience with the North Italian and Brazilian workers at Malabar-do-Brasil. There is only one conclusion to be drawn ... that any one of the foreign-born workers is worth three average American workers, whether in the category of college boys of considerable intelligence, or in the category of the migratory wandering laborer that comes to us from our own area and from the South and Middle South.

This qualification of worth is not based alone upon the capacity to work or the cheerful willingness to put in long hours during the emergencies which sometimes arise in harvest time; it is based upon such factors as the use of brains, interest in what the worker is doing, pride of achievement, and of doing a job thoroughly and well; it is based above all upon a sense of responsibility and the capacity to learn from day to day from the very jobs and tasks the worker is performing.

Out of fifteen years of experience in this country, twenty years of experience in Europe, and several years among the agricultural workers of Brazil, there have gradually emerged many basic reasons why so many American agricultural and livestock workers are inferior to the workers of many other nationalities. Perhaps the most important and fundamental is that life in the United States is too easy for young people and that there are families, in moderate or even poverty-stricken circumstances, which fail somehow to provide their children with a sense of responsibility and a pride in achievement, which in turn are the very fundamentals of any career that is successful or of any value to the country or to one's fellow citizens. Too many American parents expect the schoolteacher to provide the fundamental education and the basis of character and achievement for their children. These lacks in fundamental training produce young people, and especially young men, who are singularly immature and frequently even at eighteen or twenty seem more like children than mature men. I would say that as a rule the average European boy of eighteen is as mature, as dependable, as sober, as thorough, and as responsible as the average American of twenty-five or thirty.

Again, I think there are many factors contributing to this immaturity. They begin in the American home, where all too often there is never any really intelligent conversation around the table, but talk limited largely to gossip or baseball or golf or how the old man's business is doing; and home is the very beginning of education. Without a base established in the home, the schoolteacher and professor have little ground to work upon. Even the average knowledge and interests of countless college and university graduates of mature age are shockingly deficient; one wonders sometimes how they ever obtained a college degree.

And to be sure, too many young people look upon four years of college or the university as a kind of lengthy holiday during which their parents provide them with a car and plenty of money to spend. All too often those four years of college are merely a prolongation of adolescence, in which the virtues of maturity and responsibility do not develop at all. I judge all this from a much wider contact with middle-class, college-bred American citizens than most men, and from their sons who come in summer to work at Malabar. Most of this younger generation has been attractive, intelligent, often charming and very pleasant to have around, but when it comes to maturity, to responsibility, to satisfaction in achievement, one might as well be associating with delightful and rather precocious children.[5]

Many of these young fellows ... indeed the majority ... come to Malabar because they are in agricultural college or because they are trying to make up their minds as to whether they want to take up agriculture as a career. They express a great interest in agriculture, but generally it develops that they are looking for a pleasant way to spend the summer or to keep in training for football or because at Malabar there is swimming, riding, fishing, and all the things which go with a summer holiday. (Those things exist at Malabar because the farm lies in a very beautiful region with forest, springs, lakes, and streams, and because they are actually a part of the farm program at Malabar and considered a part of any well-managed permanent agriculture.)

At Malabar there exists one of the largest agricultural libraries outside an agricultural college. It contains not only the standard agricultural works, but countless books and pamphlets of advanced scientific importance which are frequently unavailable in most libraries. For many years

[5]There have, of course, been notable exceptions among the young Americans who come to Malabar, but invariably, on meeting their parents, one discovers the reasons. The boys with a sense of responsibility, with enthusiasm, and with some degree of maturity and direction come from homes which are in reality *homes*, in which they are taught a sense of values and aided in discovering what it is they want to do in life. However, these young men and their parents seem to be an exception, even in the prosperous, middle-class stratum from which most of the young workers at Malabar are drawn.

now, virtually every book or pamphlet on agricultural advances, from machinery to antibiotics, in many languages, has come to funnel into the library at Malabar. Yet not one American boy in ten has ever made any effort to take advantage of the vast stores of knowledge, and especially advanced knowledge, which is available there. At Malabar there are weekly Sunday farm tours in which thousands of farmers, scientists, gardeners, *et al.* take part every year. They are conducted by the proprietor, the manager, or by one of the European boys who are with us permanently and quite as capable as myself in answering the questions which come without end; but not one American boy in ten ever takes the trouble to participate in one of the tours, which are in themselves seminars in which great stores of new information and practical ideas come from the visitors themselves, practical farmers and scientists alike.

A good many of the American boys display little or no interest in the actual practical working of the farm itself. They will cheerfully and irresponsibly perform tasks which they are asked to do, but will display little curiosity as to *why* these tasks are necessary or *why* they should be done in a certain way. And almost never do they display any pride in doing the job well or better than the next fellow. Very often they will simply ride round and round a field on a harrow or a hayrake with very little attention to the task at hand. Their minds are elsewhere, on what I have never been able to discover, although I suspect that at times their minds are engaged with nothing at all, but are merely a blank. The extreme example occurred with one young American, who was not a boy at all but twenty-four years old, who was determined (so he said) to be a farmer and was at Malabar to learn. Twice after he had been with us for two years I found him driving the forage harvester round and round the field with the back end of the self-unloading wagon attached to the harvester wide-open. For those who are not familiar with such machinery, this merely means that the forage harvester he was driving was cutting the meadow grass and blowing it, not into the wagon, but straight through it and back onto the field. He had been two or three times around a field without noticing the fact. (Needless to say there came a time when we felt we could no longer afford to go on training him to be a farmer, and I will admit that he was an extreme example.)

It cannot be said that these boys failed to display interest either in the practical operations of the farm, in doing a job well, or in taking advantage of a fine agricultural library because they were overworked with long hours, for Malabar has perhaps, because of its high degree of mechanization, the easiest working hours of any farm in the country. Rarely does any worker, outside the dairy, get into the field before nine o'clock; there is at least an hour and a half off at lunch and everyone is out of the field before six o'clock in the evening. The idea of overwork would be well exploded by the fact that a good many of the boys have finished their day's work and then spent their evenings lifting weights or practicing football.

To be sure, very few of these boys come from a farm, for in summer the average farm boy in an agricultural college has his own farm to return to, a farm where there is plenty of work to be done; and the average farm boy is much less likely to be spoiled with plenty of money and a car to take along with him. As often as not he is earning his way through college; also as a rule he is far more mature for his age than these boys who come from the great American cities and the suburban middle class, which *should* be providing the very backbone of enterprise and responsibility in the nation.[6] Sometimes, in the case of the farm boy, the shoe is on the other foot ... that college tends to corrupt him or give him ideas that somehow farmwork and the raising of cattle and hogs is beneath him and that if you go to college you must be a specialist of some kind who will tell other farmers what to do. There was a time when the specially trained agricultural graduate had a mission to perform in the field of education, and there is still a great need for such education, but the field is by now badly oversupplied, and the jobs scarce and not too well paid. The agricultural colleges themselves are largely responsible for this peculiar snobbery—that it is better to be a poorly paid bureaucrat or specialist than a sound, property-owning independent farmer—for their tendency is to seize at once upon any bright boy or girl who comes to college with the intention of being a better farmer, and to attempt immediately to convert them to some specialized task and to a snobbery toward such fundamental things as the manure spreader and the cow and the pig. One great agri-

[6]It is not perhaps without significance that the vast majority of the names listed in *Who's Who* are those of men who came from farms or from small communities in agricultural territory. Even with the immense development of industry and cities, the leadership of the country in all fields seems still to come very largely from rural areas where a sense of responsibility and maturity grows from work and contacts with the soil and with livestock, and where, more often than not, a boy or a girl learns to work and to think for what he gets in this life. Yet in all of this there is a paradox, for often enough the successful man who comes up the hard way is the one who ruins his own children's lives by removing from their existence the very elements which make for character and a sense of responsibility—all in the desire that his children should have "all the things he never had."

cultural college, I discovered, over a period of three years was sending an average of 94 percent of its graduates *away* from the farm into some specialized teaching or bureaucratic field.

The fact still remains that the good farmer is *the* fundamental of any good agriculture and that his mere example to his neighbors is more important to agriculture and the nation in general than the huge army of bureaucrats and technicians, despite all the great contributions made by the inspired and devoted members of this important group.

In the field of machinery, and especially harvesting machinery which is frequently complicated and subject to accident, it is very nearly impossible to entrust any of it but the simplest field tools to the average American boy or transient American worker. Again, they will drive round and round a field without looking behind them, without listening for the ominous sounds which tell even an amateur mechanic that something has gone wrong inside the machine which may end disastrously. The first awareness comes to him when an expensive machine tears itself apart.

Behind this perhaps lies a special psychological and material factor ... that in this country machinery is abundant and low-priced by comparison with machinery in other countries. The average American boy, quite naturally, grows up with the feeling that if you break up a machine, you can quickly replace the parts or the machine itself, or maybe even the insurance company will take the loss.[7]

To the average foreign-born boy, any expensive agricultural machine is a treasure, and something almost sacred. It is not something to be wrecked and turned in after a year or two; it is something which he respects and cherishes. He takes care of it; he listens for the faintest sounds which might indicate trouble. In 1947 we bought two new jeeps. One was given over to the then herdsman, a young American of thirty-four; the other was given into the care of a young Britisher on the farm. At the end of two years the herdsman's jeep was sold as junk for twenty-five dollars; we are still using the other jeep daily; it has been reconditioned for around a hundred dollars and is as good as new.

[7] In this respect we had recently an extraordinary example of the foolishness of American parents through an accident which occurred on the roadside at Malabar. The boy, only sixteen, came racing around a curve in a brand-new convertible at eighty miles an hour, endangering anyone who might have been on the road beyond the curve. Quite naturally he went off the road, narrowly missing a telephone pole which might have finished him off for good. He got off with a few bruises, but the car was demolished, and three days later he was provided with another brand-new convertible by a father whom one could only suspect of a desire to get rid of his son altogether.

It may be, of course, that long residence and many contacts with farmers and agriculture elsewhere in the world have given me a special and
cranky point of view, but on the other hand the same experience may
make it possible for me to see our own country and people more clearly
and to understand them more thoroughly. I think the same influences, or
lack of influences, which produce the American boys at Malabar are very
largely responsible, together with forced military service which makes it
impossible for any boy to make plans to take a permanent responsible
job, for the widespread juvenile delinquency which so troubles the
nation. It should be doubly alarming, for it is out of these young people
that our future society will be built and the character of the nation determined.

In all of this I can only say there has never been a young foreign-born
worker at Malabar whom I could not send out into the field to do a job
and know that he would do it well and thoroughly, that he would take
pride in it, respect the tractor and machinery he was using, and use his
head and judgment if a crisis arose. Where such men are in charge I could
absent myself for a period of a year or more, merely leaving instructions
behind, and know that things would go well and as I wished them. I am
unable to say as much for the great majority of American-born farm
workers, either young or mature.

All too often the psychology and attitude of much farm labor today is
sooner or later to take as much as possible while giving as little as possible. Very often the worker would be good-natured and do the work while
you were present and gold-brick the minute your back was turned. And
hardly ever is there any sense of participation in a common task or aim
or in doing a job well. Frequently farmwork is merely a way of making
enough money to have a house, a radio, a television, and a car that could
be turned in for a new one every two years. Even where profit-sharing and
actual economic participation were offered and when a little more care
and thoroughness or an hour or two a day more work was involved to
increase their incomes as far as they liked, only one goal counted; that was
the amount which would provide them with the car, the radio, the television, and the house. Beyond that they were not interested.

It is true, in considering these things, that they are not necessarily
indicative of all Americans, or indeed of countless sound and skilled
industrial workers, or of the few exceptions which have come our way in

the agricultural field. The fact is that the level of quality in farm workers has been gradually declining for many years, and very sharply during the continuous war years since 1940. This has been especially true in the industrial areas where there is almost irresistible competition from the high industrial wages of the war years which no farm, however successful, can afford to pay. Even the fact that many a worker at Malabar with good wages, with housing, fuel, light, and most of his food is much better off and able to save far more money than many a skilled industrial worker, has not much altered the situation. The only pool of good agricultural workers from which the farmer can draw today is that infinitesimally small one made up of the kind of man who *wants* to be a farmer and would not work in a factory, putting nuts on bolts for eight hours a day, no matter what the wages. We have had, and have, this kind of man, and when you can find him he is a treasure to be guarded.

The decline in the quality of farm labor has created a kind of vicious circle for the really modern farmer. The lack of respectable and responsible labor or its scarcity has forced a continuously higher degree of mechanization upon the farmer, and as mechanization has increased, the need or the use of the old-fashioned general "hired man" has become increasingly impossible. Every permanent worker today on a modern farm must be a skilled worker, capable of handling and caring for expensive machinery or valuable and sometimes registered livestock. He must be not only a man who is trained and possessed of sound knowledge in his particular field, but he must also be a responsible man. As any farmer knows, these are today few and far between. At Malabar the problem has been largely solved by the joint employment of the few reliable American-born workers we have been able to find and the European workers who have belonged universally in our own experience to the category of worker vital today to any agricultural or livestock enterprise.

Mexicans and Puerto Ricans can hoe sugar beets and migratory workers can pick fruit and harvest vegetables, but these are specialized fields more in the realm of industry than of agriculture and livestock. Many native-born Americans consider themselves above such labor as that involved in the growing of sugar beets and in the vast vegetable operations in Florida and the Southwest, and a good many of them will sit and take public relief funds rather than earn their living in such a fashion. And there is no doubt that the average Mexican or Puerto Rican is a better, more responsible and conscientious worker than the American who feels himself too good for such work.

In all the sentimentality, in all the mushy thinking regarding the plight of the well-paid American migratory worker, one point has been consistently overlooked. The great majority of them are migratory workers not through economic necessity but because they *like* to be migratory workers They like the life, and under no circumstances could they be tied down to the responsibilities and boredom of permanent citizenship in any community, large or small. On the whole they are a happy people because, unlike many a city dweller or a bad farmer, they are doing what they want. They will move north with the spring and summer in a car and sometimes with a trailer until at last autumn finds them somewhere near the Canadian border, when they return to Florida or the Rio Grande to begin the cycle all over again. In fairness to them it must be admitted that they are an invaluable factor in our horticulture, agriculture, and economics, without which the vast business of feeding the American people on the highest dietary level in the world and with the best and the cheapest food in the world would be impossible.

I am quite certain that almost every farmer will understand what I have been writing about, although much of it will be new and perhaps startling to the average city dweller. The question of good farm labor today is one of the farmer's most serious problems, and happy and fortunate is the farmer who has plenty of sons and daughters who can work with him and possess maturity, intelligence, and a sense of responsibility.

This lack of good and responsible labor is one of the great handicaps of American agriculture, especially in any large or intensive operation. It may be that when employment and inflation level off it will be possible once more to find good men in quantity who want to work well and with a sense of pride in achievement and with the freedom, the health, and security which go with country living. Much of what I am writing may seem harsh, but I cannot help believing that it is true. The lack of good farm labor has blocked us at Malabar again and again from doing what we wanted to do and from achieving many things which "can't be done."

The situation has driven us more and more into the most intensive possible mechanization, with fewer and fewer irresponsible or lackadaisical workers, and has forced us to exercise our ingenuity and imagination to cut down the number of operations necessary to fit and cultivate and plant a field, to feed and milk a cow, to feed and farrow pigs, to maintain a good pasture. It was this necessity which drove us into many improvements and many devices which otherwise might never have been undertaken. Some of these were in the realms of what "can't be done."

Although we raise at Malabar a variety of livestock and of crops, sometimes only for the purpose of experiment or research, we are essentially a grass farm with about three hundred acres of grass and legumes as our basic crop. The making of three or four cuttings a year of hay and silage on such an acreage, as any farmer knows, is quite a job, indeed an almost heroic task, not only because of the great tonnage, but because it must all be cut almost at the same time while the hay is in prime condition and has not become overripe, woody, and lost much of its feeding value.

The problem of the first cutting, always of greater bulk but of inferior quality to later cuttings, especially as our meadow mixtures consist of alfalfa, brome grass, and ladino clover, was largely solved years ago by making it into silage. This process permitted us to cut labor costs because it could be highly mechanized with a forage harvester and self-unloading wagons. The principal labor problem came at the upright silos where it required extra men, and unloading was a slow process, menaced always by the plugging of the blower pipes. Later on during the winter feeding season extra time and labor were needed in throwing down the silage and carrying it to the cattle. Many elaborate upright silos were and are offered on the market but many of them are beyond the means of the average practical farmer who must figure how many years of milk or meat it will take to pay for the installation. Some of these silos offer mechanical unloading devices and some of these are very effective, but at Malabar we have kept as far away as we could from mechanical devices, which by breaking down could block all operations for long periods of time. The alternative to all of this was obviously the trench silo in which the blower, with its handicaps of slow unloading and plugging, was eliminated. The forage harvester could cut and blow the silage into a self-unloading wagon and all that was necessary was to dump it into the trench silo. The feeding of the silage later in the year was also greatly simplified and cut labor and expense more than half. So, together with many another farmer, we found trench silos indispensable in a program of grass farming in which very heavy tonnages and large acreages are involved.

Our first trench silo was excavated nearly ten years ago and since then we have made many experiments and finally arrived at an operation in which labor is reduced to a minimum and efficiency is raised close to the maximum. The final result of the experiments is a five-hundred-ton trench silo built into the side of a hill which can be filled from the top merely by backing up to it the self-unloading wagon and discharging its load in a matter of three to five minutes. A Ferguson tractor, equipped with Bombardier half-tracks and a scraper and operated by one man, levels off the silage and packs it.

Later in the year the silage can be self-fed through a movable rack which is moved back as the silage is consumed, or it can be put in nearby bunks by a Ferguson tractor equipped with a high lift capable of lifting a thousand pounds at a time. In the latter case one man can feed sixty cows for the day in half an hour or forty-five minutes. The silage is uniformly excellent in quality and the spoilage is negligible, both in quantity and in relation to the immense saving of time and labor.[8]

In the course of experimenting on silos we discovered that a roof or covering of some kind is advisable over a large trench silo because the surface exposed to weather and spoilage is considerably greater than in the case of the upright silo, and quite by accident, in the course of digging a trench silo beneath a shed for which we had no use, we discovered that after the silo was filled we had a large storage space left beneath the shed roof. This we promptly utilized for the storage of baled hay and straw. Since then similar shed roofs have been constructed over all trench silos and so have provided us under one roof with storage for great quantities of silage as well as straw bedding or baled hay. The bales are stored imme-

[8]In making the silage there is added to it about ten pounds of salt, five gallons of molasses, and sixty pounds of whole grain barley, wheat, or ground corn to the ton. This makes an extremely palatable silage, easily able to compete with corn silage in attracting the taste of the cow or steer. The salt, as in sauerkraut, cuts the acid bitterness, while the molasses provides some nutrition and greatly increases palatability. The small grain or ground corn ferments and turns to brewers' or distillers' grain, which every farmer knows is the best of all livestock feeds. Barley especially is excellent in the mixture. The grain becomes soft and milky and being mixed *with* the roughage is acted upon by the cow's stomach, which is not made for the consumption and digestion of dry grain. Mixed with the roughage, it is consumed with a much higher degree of efficiency than is dry ground grain. In the case of wheat and barley no grinding is necessary and this tiresome and time- and labor-consuming operation is wholly eliminated. In other words this silage mixture provides a *whole* ration of excellent quality, giving fine results with dairy cattle, young stock, and beef cattle. The feed value, made up of high-protein alfalfa-brome-ladino mixture, plus molasses and the starches, as well as the protein in the grains, is extremely high. But what is vital here in this discussion, is that the process eliminates hundreds of work hours each year.

diately after the silo is filled and piled directly on top of the silage so that their weight further packs the silage and serves to shut out the air, thus cutting down considerably the wastage involved. At Malabar, however, we do not consider spoiled silage as waste. Either it is mixed into the soils of the fields in a process of sheet composting or added to the compost heap where, in the high-production truck garden operations, it has more value than when used as feed.

Although we are great believers in the virtues of silage, we do not overlook the place of good, high-quality second and third cutting hay, especially when it is made of such a mixture as alfalfa, brome, and ladino clover. A forage harvester could have made these later cuttings into chopped hay, but without using an artificial drying process, we found ourselves unable to make the quality of hay desired with a forage harvester. If we took the hay into the farm in a condition which we considered green enough, it heated so that it threatened to set fire to the barn or turned brown and lost much of its feed value; if it was dried sufficiently in the field, the leaves blew off both in the field and in the mow and we were left with little but the stems which contained low feed values. The only alternatives were the construction of ducts and flues for pushing air through the hay to dry it after it was in the barn, or an expensive heat-drying process which involved more labor and was too expensive for the average farmer as well as for ourselves. So in order to get the high quality of hay desired, we were forced into the much more laborious process of field curing and baling. We found that we could bring hay that had been cured and baled intelligently into the barn quite safely with a much higher moisture content and greenness than was possible with chopped hay or even loose whole hay. The bales, if stacked properly, would dry out in the barn, leaving us with hay that was dark green and very near to alfalfa leaf meal in quality.

Most of this hay was stored in an enormous old-fashioned barn with loose floor boards and siding which admitted plenty of air, and through this circumstance and the existence in one end of the barn of a large squirrel cage fan, we made an accidental discovery which promised to solve all our haymaking difficulties and cut labor costs as much as 75 percent.

It happened on one occasion that, in the face of continuing bad weather, we brought in some baled hay that was far too high in moisture to be safe from heating and molding and the idea occurred to me of turning on the big squirrel cage fan in order to suck currents of air through the sides and floor of the big barn. The plan would also give us a second valuable use for the fan, which had been installed originally to pull the steamy, damp air from the big loafing shed below during the winter season. The plan worked almost perfectly, for the fan brought in through all the cracks and openings strong currents of hot, dry summer air, much of which flowed through and over the loosely piled bales. The air currents served two purposes: (1) actually to dry the hay and (2) to eliminate constantly the damp air that arose from the bales. In other words, the air circulating in and around the bales was always dry and the damp air was constantly being removed.

Out of this experience came the hay-drying pilot plant erected at Malabar during the summer of 1954, which promises to greatly increase and stabilize the quality of our hay, to make us independent of the weather, and to cut enormously the time and labor of making good hay.

The pilot building, provided for the experiment by the Reynolds Aluminum Company, with which Malabar has worked closely for many years, was simply a standardized machine shed manufactured by the company, of pole construction and covered with sheet aluminum.

The important thing was the principle involved, which so far as I know was employed at Malabar for the first time ... that of *pulling* air through hay to dry it rather than attempting to push it through in a system which was nearly always clumsy and inefficient.

The building itself was quickly and easily built, and made as airtight as possible save for the floor which was raised two feet off the ground and simply covered with old fence wire that provides the minimum of resistance to the passage of air. At the top on one side we installed a five horsepower electric fan, made an opening with a sliding door on the opposite side through which we could blow the chopped hay, and we were ready to go. In principle, once we covered the wire mesh floor with two or three feet of chopped green hay and turned on the fan the hot summer air *had* to come through the hay. It is interesting to note that although we have not yet found anyone anywhere who has made a try of pulling air through, we were assured on all sides that it could not possibly work.

Needless to say, the aluminum structure, without windows or doors, created considerable excitement among neighbors and visitors who were puzzled as to its use, and those of us at Malabar awaited impatiently its completion so that we could give it a thorough test.

We put in the first chopped hay the moment the second cutting was ready, and the first test was an acid one, for in our impatience we cut the alfalfa when the very first blossoms were showing and the moisture content was at its peak. In addition to this, the hay still was moist from the rain of the night before. We cut direct from the field using the forage harvester as a mower, blew the chopped green stuff into a self-unloading wagon and thence into the drier itself. After four or five heavy loads, we were ready to go, and turned on the big fan. ... The air not only came freely through the hay but the current actually tended to lift it into the air.

In that first test under the most extreme conditions, it required about twenty-four hours to dry out the hay. The resulting product was deep green and almost silky in quality.

Test number two was made with *whole* green hay cut direct from the field with the forage harvester. In this operation all the knives were taken off the harvester, which acted simply as a mower with a blowing attachment which carried the hay into the self-unloading wagons. Hay cut whole in this fashion and blown directly on top of the earlier dried, chopped hay dried in about half the time and with the same high quality and fineness of texture.

The third test was made with hay which had been mowed for field drying and baling and was only partly dried when it was caught by a heavy rain. This was picked up out of the windrow by a forage harvester with a pickup attachment and put into the drier. It too dried within twelve hours and with no perceptible difference in color or quality. If we had attempted to complete the drying of this hay in the field, the result would have been discolored and low-quality hay. By putting it into the drier while still wet we managed to make of it excellent deep green forage of high quality.

The fourth test was made by mowing down whole hay in the field and permitting it to wilt and dry in the sun to a point where approximately a third of the moisture was removed. It was partly dried but not sufficiently for any shattering of leaves to occur. It was then windrowed and picked up by the forage harvester and blown into the drier, where only about six hours were required to dry it. The results were equally good in quality and texture.

It was this final test which we settled upon as the most practical for most farmers and one which we shall probably use so long as we have but one hay drier. It requires two more operations (the mowing and the

windrowing for the pickup attachment) but these operations can be done
in turn by a single man and the speeding up in the drying process itself is
an advantage. In other words, by partial drying in the field, more than
half of the drying time required for the chopped green hay cut directly
from the field and put into the drier, is eliminated.

In the eventual drying operations at Malabar, the slower drying will
not be a disadvantage as we expect to put up a drier of the Malabar-
Reynolds model at each barn, which means a total of three. It will then be
possible to partially fill one drier, turn on the fans, move on to the second
drier, repeat the operation, and finally to the third. By the time we have
put in four to five feet of green hay in the third drier the hay in the first
will be well dried and we shall return to it and repeat the process so that
the operation will be a continuous one until all three driers are filled. In
experimenting, we discovered that it was possible to put in green hay to
the point of impaction where the weight of the *green* undried hay in set-
tling began to close the spaces which permitted free passage of the air
through it. This, in the case of chopped hay, amounted to a mass about
three to four feet deep. With the whole hay, unchopped, it was possible to
put in almost twice as much before we arrived at the point where the hay
tended to pack and block passage of the air.

At the very beginning of the construction a second idea, almost as
important and valuable as the drying process itself, occurred to us. This
was to build in doors approximately six by four feet, which during the
winter months could be lifted up and fastened and permit the cattle to
feed themselves. At the same time the roof itself was extended out ten feet
on both sides of the drier to provide a shelter and to prevent the space
where the cattle fed while eating from becoming a mudhole. The doors
were so placed that the feeding level came across the chests of the cattle;
thus they were able to reach into the drier and feed without wasting hay
by dropping it on the ground. In all but the dairy barn there was no waste
whatever as hogs are run with cattle and eagerly pick up any of the dark
green alfalfa that falls to the ground. At the time of writing we are half
through the winter and both cattle and hogs have been fed from this self-
feeding drier with no troubles of any kind.

Of all the projects at Malabar none has been so spectacularly effective
in reducing costs in time, in labor, and in gasoline as this hay-drying
operation.

In the pilot plant, which was simply a standardized machine shed, there were many "bugs" which will be eliminated in future construction. One was the leakage of air since the building was not completely airtight and much air came into it which did not pass through the hay. It is probable that an airtight structure will reduce the drying time all along the line by an hour or two. The dimensions of the present drier are twenty-six by thirty feet at the ground level and eighteen feet high at the roof peak. Future models will be twice as long and half as wide (that is to say, sixty feet long and thirteen feet wide) and will include the installation of two five-horsepower fans rather than one. It will also have a shed rather than a peaked roof, which will permit us to do away with the supporting cross construction in the roof peak, which interferes on the inside of the drier with the distribution of hay from the blower.

The savings in time and labor over the making of baled hay are truly enormous. In the process of making baled hay of good quality in the field, several operations are necessary ... the mowing, two or three or more times over the field to windrow the hay, the baling, the loading, the transference of the hay into the mows, and finally the carrying of it to cattle for feeding during the winter months. To accomplish all this in terms of three hundred acres of hay all ready to be cut at approximately the same time, a whole crew of men is required, especially if the weather is uncertain.

As against all of these operations, the use of the drier reduces the whole thing to what is in reality almost a single operation with crews of three or four men ... one for the forage harvester, and two or three for the wagons.

Where the hay is cut direct from the field, all the operations required in the field for making baled hay are eliminated, save that of cutting the hay and blowing it into the wagon. In winter, operating the drier as a self-feeder, there is no feeding labor whatever save perhaps at the end of the season to push forward what hay remains to bring it within reach of the cattle. In other words, throughout the whole operation from field to manure pile the hay is never touched by hand.

In addition to all this, the risk of spoilage and loss from bad weather, as in the case of grass silage, is wholly eliminated, so that the combination of silo and drier assures the farmer of 100 percent top-quality forage of a quality equal to alfalfa leaf meal. The tests made on hay from the Malabar-Reynolds type drier by the U.S. Agricultural Station at Beltsville,

Maryland, showed that it contained approximately one-fifth more protein and carotene than the best field-dried hay and a much wider margin over much of the hay dried on the principle of *pushing* air through hay. The product had virtually the values of alfalfa meal and a vitamin test run on this naturally dried hay may well show a margin *above* alfalfa meal, since in most of the processes used for making alfalfa leaf meal heat at very high temperatures is employed which can be very damaging to certain vitamins.

The greatest virtue the drier set up at Malabar is perhaps its simplicity and low cost. The costs on the building itself constructed by outside labor at Malabar amounted to about sixteen hundred dollars complete, which is approximately the cost of our baler twine bill for one year. It is of pole construction with the frame and aluminum siding fastened to treated poles set in concrete ... a construction which involves almost no costs for repairs and upkeep and will last for generations. It can be constructed by any farmer with the slightest knack for carpentry at a much lower cost. And the cost of the electric current to operate the fans is extremely reasonable and insignificant in comparison with the labor and operational costs of hay made and fed in other ways.[9]

For a farm smaller than Malabar, the drier takes on an even greater importance, since actually the whole hay-drying operation can become a one-man affair, with the farmer himself driving the forage harvester and the self-unloading wagon and operating the blower. In fact, most of the hay put into the drier at Malabar was the result of such a one-man operation. One of our men, Hasse Lindgren, was responsible for filling the drier while the rest of us went about other tasks. The pilot plant drier at Malabar, when filled, contains between thirty-five to forty tons of dried hay, but such a drier, of completely simple construction, can be made in any size, provided that there is sufficient fan power to pull the air through the hay.

With such a drier, a hog or poultry farmer can make his own alfalfa leaf meal at a fraction of the money he spends in buying it, and high-quality hay from such a drier serves as a rich protein–carotene supplement for dairy cows, young stock, and feeder cattle. In fact, if fed in unlimited quantities, such hay is likely to produce diarrhea in mature cattle and scours in young stock, and some control over the feeding of it

[9] [Complete and detailed plans for the construction and operation of this Malabar-Reynolds hay drier are available from the Reynolds Aluminum Company, Parts Division, Louisville, Kentucky.]

must be exercised.

We are great believers in grass silage or silage of any kind for livestock, but we do not overlook the equally great value of good hay, for all hay is in fact a concentrate of field grasses and legumes with about 75 percent water content in weight removed. The better the hay, the higher in protein and carotene concentrate. We have put cattle herds through the winter very successfully on a diet of silage alone, and the late Ed Babcock at the time of his death had high-producing dairy cows at Sunny Gables Farm which had never known hay and been fed throughout their lifetimes on silage. We have found, however, that excellent hay, and especially of the quality obtained by the drier, produces splendid results in milk production, growth, and in the feeding out of cattle, not only because it is in itself a concentrate but because it contributes greatly to the capacity of these same animals to assimilate and utilize grains and other feeds. As against silage of any good quality, top-quality alfalfa mixture hay has a concentrate feeding value of at least three or four to one.

The use of the *pulling* rather than the *pushing* principle in drying hay quickly and efficiently is very largely new and when we attempted it we were told in many quarters that it could not be done. We were not, however, discouraged for we were backed by the simplest principles of physics—a fact which has apparently gone unnoticed throughout all the experimentation in hay drying. In *pushing* air through hay, the velocity becomes constantly lower in direct ratio to the distance from the fan and as it loses force, it can become blocked at the very place where the hay is greenest and highest in moisture content.

In *pulling* air through hay the contrary is the result. In other words, the velocity of the air current tends to increase as it approaches the fan and at the same time the greenest and dampest portions of the hay. In an airtight drier, partly filled with hay, and with only the bottom open for the air to come through, the powerful fan sucks out all air from above the hay and the only air to replace that which has been sucked out by the fans *must* come through the hay. I am told by one agricultural engineer (who said the drier could not work and continued to say so even when it was filled with dried and perfect hay) that I must not call this absence of air in the upper part of the drier a vacuum; but if it is not a vacuum I do not know what it is. And I will not be technical about it beyond observing that one of the simple rules of physics is that nature abhors a vacuum and that, whether it can or can't be done, we are very satisfied with the whole operation because it works, even beyond our expectations.

While in the hay drier we achieved what couldn't be done, other undertakings at Malabar have not always been so successful. Some have failed partly and now and then we have had a total failure. But both successes and failures have tended to make life interesting and exciting, and I doubt that we have ever begrudged the time or money spent upon experimentation and research. Many times we have gained in the experience and acquired knowledge which proved useful in other fields.

Perhaps the most interesting large experiment which we ever attempted was the growing of corn directly in meadow sods. Corn, even when the production per acre is large, is an expensive crop, and on countless farms today the harvest does not pay for the expenditure in labor, time, gasoline, taxes, interest, and general wear and tear on the soil and the machinery. With all this in mind we have been searching constantly for the means of getting good yields of corn at costs greatly below those of the average corn-growing operation. At Malabar we always have meadows of heavy alfalfa, brome grass, and ladino clover available for experimenting. On one occasion through an error the amount of ladino clover seed contained in the mixture planted was much higher than we usually sow, with the result that in the first summer the heavy ladino seeding grew so rapidly that it virtually choked out all the alfalfa and brome grass seedlings, leaving us with a lush and solid field of ladino which was of little use. It is virtually impossible to make ladino clover into good field-dried hay, and under many conditions ladino creates an actual menace of bloat if used for pasturing cattle. As poultry and hog pasture nothing is more valuable, but we already had plenty of pasture for the few hogs and chickens we carried at that time.

I think it was Dr. Roger Bray of Illinois University who gave us the idea, uttered half in jest, of using the field to grow corn. We were all agreed that while it was an excellent practice to plant cover crops with the last cultivation of corn, it was rare indeed that the cover crop amounted to anything. "Why not," suggested the good doctor, "plant corn in a *permanent* cover crop?"

The suggestion set us off. Corn, we knew, belonged to the family of grasses. Its moisture and fertility demands were high. In South America during the summer season where there was abundant rainfall I had frequently seen excellent corn growing in the face of the heaviest competition from weeds and other grasses.[10]

The plot chosen for the experiment was almost solid ladino with here and there a patch of alfalfa and a few tufts of brome grass and volunteer orchard grass. Our plan was to cut narrow strips through this sod and plant the corn in the strips and simply let it grow without further cultivation or attention. So eager were we to give the experiment a try that we set to work at once although it was already late in the season for planting corn. No machinery existed for such an operation so the big Seaman tiller was rigged up by removing part of the blades so that it would cut two strips about ten inches wide and nine inches deep through the sod. The strips were marked out by putting the fertilizer on in advance with a corn planter. These strips of fertilizer were then chewed up with the Seaman tiller and in doing this we achieved the aim of any good corn farmer ... that of mixing the fertilizer into a small area *surrounding* the corn plant, rather than simply dropping it on the surface or placing it at plow-sole depth at the bottom of the plowed furrow. This operation we followed with the corn planter which put in the seed. All of the plot save one test strip was mowed and the green stuff left on the field as a mulch.

From then on we watched the plot daily and sometimes twice a day, but we ran into bad luck and bad judgment. The bad luck lay in the fact that we ran into a long period of dry weather; the bad judgment was in attempting to plant corn too late, after June 1st, and into hot and sometimes dry weather. That first experiment was largely a failure and the corn that did flourish was put into the silos. Oddly enough the best corn was on the test strip where we had *not* mowed down the grass and legumes.

During the next winter the Seaman Tiller Company constructed for us a small power tillage machine working from a Ferguson tractor which did perfectly the job of cutting strips through the sod. The same procedure as before was followed. This second year we got the corn in during the second week of May and we had good rains and immediately there was a

[10]In Brazil it is a habit to plant corn on newly cleared land or land which has been in eucalyptus. Frequently the corn is planted on unbelievably steep slopes from which it would be possible to harvest the corn simply by picking it and throwing it to the bottom of the hill. Yet little or no erosion occurs even under the heavy tropical rains. There are two reasons for this. One is that the top level of soil contains a great amount of leaves and all sorts of rubbish and so drinks up the rain. The other is that much of this corn is given at most one cultivation or none at all, and under the hot sun and heavy rains the ground is quickly and spontaneously covered by a rank growth of weeds and grasses. In other words, the heavy rains themselves plus abundant sunlight save the Brazilian countryside from an erosion which under other conditions could be devastating.

startling difference in the experiment. The corn, which was closely planted, six to seven inches apart in the row, came on rapidly and within six
weeks had virtually shut out the sunlight. From the moment of planting
it forged ahead and at the end of the season we had a crop of corn on
which the yield was about twice the average yield of the state and everybody was pleased. We had raised an excellent crop of corn with at least 80
percent less expenditure in time, labor, and gasoline than was needed in
the accepted, conventional way of raising corn.

But everything in agriculture is extremely complicated, and one test
experiment is rarely enough to establish a sound judgment, as we discovered the following year. In this third year the corn was planted early and
started off well, but there was a lack of rain and it began to appear that
the experience of the first year's planting would be repeated. The crop was
somewhat better than the first ill-fated year but it was still a poor crop.
Then the fourth year we had an excellent crop and took off the plot corn
silage at the rate of twenty-three tons per acre, which is very high. The following year there was again a shortage of rain in the early season and
again we had a poor crop. But the important fact was that in the whole
experiment we had reduced the cost of raising corn by a really stupendous margin; we had eliminated all erosion and water runoff and mini-
mized almost to nothing the destruction of organic material, which is one
of the worst debits to be raised against corn.

In those five years we discovered we very nearly fixed the factor of success or failure upon abundant rains at the right time, which was early in
the season. If we had these rains, the corn was a great success, even
though there was a shortage of rain later in the season. It was very clear
that the critical time in the experiment was the first six weeks. When there
was plenty of rain the young corn shot up rapidly, and because it was
closely planted, it shut out the sun from the sod in which the corn was
growing and from then on the corn had the advantage. The grasses and
legumes went backward and frequently died out altogether and the corn
shot ahead. It was also clear that once it got a flying start, the corn robbed
the grass and legumes of moisture rather than the other way around. If
the corn got a slow start, and the weather was dry, the already well-estab-
lished root systems of the grasses and legumes robbed the corn and from
then on the situation grew progressively worse.

We also discovered that we had been wrong to clip the field with a mower before we planted, whether we removed the resulting hay or left it as a mulch. Close observation led us to the belief that where the meadow vegetation was allowed to stand, it was already by the second week in May well on its way to maturity and the development of seed and its moisture requirements became less and less each day. Where we clipped the field the grasses and legumes set up a whole new rapid growth for which the moisture requirements were very high, with the result that the grasses and legumes with their long-established root systems stole very nearly all the moisture from the corn, which was only beginning to establish a deep root system.

But there were other observations of great interest concerning the behavior of the grass and legume mixture. Perhaps the most interesting was that in the second year, when the experiment was very successful and we had a fine yield, the corn absolutely eliminated all the alfalfa and ladino clover and a few other scattered legumes of spontaneous origin. It would almost be the truth to say that not one legume survived. Two elements were involved here: (1) The corn had gotten off to a flying start and robbed the legumes of moisture, which was particularly disastrous to the shallow-rooted ladino clover. (2) What was even more important, the corn, after the first six weeks, had shut out the sun completely and so upset the metabolism of the legumes that the whole system of photosynthesis—the mysterious process by which plants make 95 percent or more of their growth from sunlight, air, and water—simply collapsed and the plants died. The curious thing was that the grasses, notably the brome and the volunteer orchard grass, survived, but in a sickly condition (and when you can make orchard grass look sick, you have really accomplished something). Here and there Italian rye grass appeared and even produced seed under the adverse conditions, although no rye grass had ever been seeded either in the plot or in any nearby field. In the third year when the rains were short and the corn crop was poor, the grasses took over completely, and we ended the season with a tight sod of mixed brome, orchard, and rye grasses and no legumes whatever. In the fourth year when the rains were good and we raised a really big crop of silage corn, the brome and the orchard grass virtually disappeared and we were left with an almost solid sod of rye grass.[11]

[11]The behavior of the grasses during the experiment proved again that the rye grasses are among the toughest and most persistent of all grasses and probably provide the best cover crop that can be found for seeding in corn at the time of the last cultivation. Rye grass, like Kentucky fescue, will grow on almost any kind of soil, including the bare clay banks

Although the five years of experimenting proved to us that growing corn in sod was a gamble and perhaps too great a gamble for the average farmer, we have not abandoned the experiment, principally because if a sound method of achieving the end can be discovered and established, the system could be revolutionary in many ways for the whole of the Corn Belt area. I will list the advantages which certainly proved themselves during at least two of the five years of the experiment.

(1) There was no plowing or fitting of any kind save for cutting the strips in which the corn was planted through the heavy sods. After the corn was planted there was no cultivation, and consequently no destruction and loss of organic material. (Two agricultural machinery companies have in cold storage designs for a machine which will cut the strips, put in the fertilizer, and plant the corn, all in a single operation, which would be a great improvement and saving over the makeshift machinery operations used in the experiments. It would mean that the farmer could plant his corn by driving across a field once and then simply come back and harvest his crop.) (2) With corn grown in sod the erosion factor and that of runoff water are totally eliminated. (3) On fields where the corn was grown in this fashion the considerable destruction of organic material resulting from heavy fitting and cultivation would be eliminated and the fields over a period of years would show only a constant gain rather than a constant loss of organic material. In other words, under such a system, one could grow corn up and down steep slopes without danger of erosion and could grow corn on the same field for generations with only a gain rather than a loss of fertility, provided sufficient amounts of commercial fertilizer were used and the strips in which the corn was grown were alternated year after year.

---

of a roadside cut, but like fescue, it is not very palatable to cattle, who will leave both grasses to the very last in any pasture. Both grasses were oversold when they were first advocated and came into use. Their best use is perhaps in the role of "recovery" grasses. Through their capacity to grow in any kind of soil where there is sufficient rainfall, they can play a valuable role in the restoration of soils and the production of topsoil. After a few years they can be plowed up and better grasses and also legumes can be seeded in to provide the next step in restoring depleted lands or in creating topsoil where there was none. We used Italian rye grass only one season in a field at the upper edge of Malabar, but like orchard grass and fescue 31, and even timothy, rye grasses have a way of traveling and now Italian rye grass turns up unexpectedly in every field on a thousand acres. One of the most interesting experiments I know is that carried on near Middleburg, Virginia, into the palatability of various pasture grasses. In these experiments, orchard grass and brome grass on well-balanced soils headed the list for palatability, and rye grass and Kentucky fescue 31 came at the bottom. Of course a pasture consisting of a highly varied mixture of grass and legumes still has the greatest feed values.

Although the experiment to reduce the high costs, both in expenditure of labor and money and in soils damage, of raising corn was a partial failure, we have succeeded in Malabar in working out a system of raising corn which very nearly cuts in half the costs and greatly reduces soil losses. The system involves the whole plan of sheet composting described in detail later in this book, as well as two comparatively new and modern implements and the abandonment of the moldboard plow in the whole operation.

The rotation at Malabar is a long one, with fields kept in alfalfa–brome-ladino mixtures for from five to seven years. At the end of this time, the fields, by this time saturated with nitrogen and the deep, heavy root systems of grasses and legumes, are ideal for the production of corn and are put to two successive years of corn production. In the first year the heavy sods, after the last cutting of hay, are chewed up with the Seaman power tillage machine to make easier their deep tillage later with the Graham plow. Then, at any time during the slack season of work in the fall, the fields are ripped deeply with the Graham plow, to a minimum depth of fifteen to eighteen inches, and left to lie over the winter. Because of the deep tillage and the trenches left by the Graham plow, absolutely no erosion occurs and all rainfall and melted snow go directly into the deeper levels of the soil. During the winter the frost action, which in these ripped-up sods will penetrate to a depth of eighteen inches even during a mild winter, does all the fitting, far better than any amount of machinery can do it, and in the spring the soil is in such condition that one could sow lettuce or radishes in it. The field thus treated is permitted to lie until corn-planting time when all the surface weed seeds have germinated and some of the weeds are already several inches high. At corn-planting time these weeds simply become green manure when they are killed and worked into the soil by heavy disks which serve to level the field for planting at the same time. After the corn is planted it is given but one cultivation instead of three. In corn planted as thickly as we plant it in Malabar the corn itself in an ordinary year chokes back the weeds by its own rapid growth and by shutting out the sun altogether. In wet years there are occasionally a few weeds but not enough to merit the expense of cultivating the whole field a second time.[12]

---

[12]Thick planting of corn is practiced at Malabar because the fertility is at a high level and because we are interested in producing not the biggest ear but the biggest number of ears and the biggest possible tonnage per acre. The same is true whether the corn is used for silage or is picked and the remaining fodder is chopped to incorporate deeply into the soil as organic material.

When the corn is picked, the Seaman tiller is put into the field to chew the fodder into small pieces which are then mixed into the soil by the ripping and mixing action of the Graham plow, an operation which can be done at any time during the slack months of late fall and early winter. From there on the same process is followed in putting in the second year of corn. This process, by opening up the soils deeply and mixing either the old sods or the chewed-up corn fodder deeply into the soil, actually makes for an increase rather than a decrease of fertility and soil structure quality and the decline which follows the old conventional ways of preparing cornland. Pictures of corn grown in this way are contained in the photographic section of this book and speak for themselves.[13]

Under this system we are able to cut the costs in the production of corn by almost 50 percent (excluding the element of commercial fertilizer, which is variable and on which the comparative costs are difficult to compute), reduce greatly the factors of erosion and depletion of organic material, and reduce to a considerable degree the effects of constantly running heavy machinery over the soils.

In all of these experiments, whether they have failed or have succeeded, we have been working steadily toward a goal which, put in a few words, is the reduction of labor, machinery, and other costs in agriculture, while obtaining optimum yields and maintaining and even increasing the fertility of the soils. Such an aim seems to me to state a perfect goal in agriculture and one which eventually, perhaps generations hence, must be realized if a world in which the annual population gains are prodigious, is to be fed. But the goal has a secondary and more immediate importance to the ordinary, individual farmer as a means of raising his profits by intensive production per acre at the lowest possible costs. Automatically such an operation increases his profits, lowers his costs,

[13]Many people and even many farmers have the erroneous idea that corn is a damaging crop because it removes a great amount of mineral fertility from the soil. Although both corn and cotton are among the greedier crops, no plant takes as much as 5 percent of its growth *from the soil*. The remaining growth is made through the process of photosynthesis out of sunlight, air, and water but—and this is a big but—the photosynthesis cannot operate properly nor the corn make a healthy productive growth if the soil in which it is growing is not of good texture, with abundant organic material, and with a proper supply and balance of *available* minerals and nitrogen. The real damage caused by corn and cotton in soils comes from the factor of erosion by both wind and water, by the constant plowing, fitting, and stirring of the soils arising from many cultivations, which in turn rapidly destroy organic materials; also by the constant use on the soils of heavy machinery. If these factors can be reduced as much as possible while at the same time reducing the actual costs in time, labor, and gasoline, a great step forward will be made in the whole field of corn and cotton production.

and establishes his prosperity upon a sound, independent, and permanent base.

Whether we fail or not in experiments made with the above goal in mind, we shall not be kept from continuing the program because someone or many someones have said, "It can't be done." To date the someones have been wrong far more often than we have been in all the Malabar operations.

## GARDENS AND LANDSCAPES

ONE DOES NOT SIMPLY PLAN, LAY OUT, AND ESTABLISH FULL-BLOWN A garden or a landscape. If it is to be properly your own garden or your own landscape, viewed from your bedroom window, you cannot hire it done. By doing so you may create a spectacular show of flowers and shrubs or a handsome, somewhat artificial landscape, but it will not be your own. Both may attract visitors who admire gardens and landscapes and who find both very fine examples, but they are likely to go away with a sense of something missing or a hidden sense of deadly familiarity on which they cannot quite place their fingers. Magnificent, but they have a sense of having seen it all before. Perhaps because frequently it resembles those artificial, flower-stuffed gardens moved in wholesale to flower shows from the estates of Long Island millionaires.

A garden, a landscape, or even a whole farm, if it is to be successful by any standard, is essentially a creation and an expression of an individual or at least of two or three individuals who feel alike toward it, who share the same aims and traditions; for tradition has much to do with the beautiful garden, landscape, or farm. There should be a *rightness* in relation to the whole landscape, to the climate, to the country, to the regional architecture, to the type of soils, even perhaps to the existence of the natural birds and wildlife. It should have a relation to the past of the region, to history itself.

When we first set out to work on the gardens and the landscape of Malabar Farm, we inherited little but the natural wild beauty of the valley, the forest, and the little stream that fed the marshes which later were to become known as the jungle. It was a frame and foundation but little else, for the areas around the various houses and farm buildings, save only at the Anson Place where old Mrs. Anson had constructed with infinite pains and labor a handsome rockery, were neglected and abandoned, shaggy and unkempt. A great many of the buildings were simply falling into ruin, and leaned at crazy angles, waiting only for someone to give them a push or set a match to them. Here and there were piles of discarded rubbish and worn-out, rusting machinery, truckload after truckload of it. The only effort at order or planning was the annual cutting down of every bit of wild growth surrounding one or two small ponds and along the roadside, an action caused, I think, simply by some atavistic fear inherited by many in our countryside of a devouring forest which took over every small clearing unless one maintained a constant vigilance.

Our Ohio forest, which, if left alone, becomes thick and tropical and impenetrable during the summer, quite naturally takes over even today any field which is abandoned for a few years. The early settlers, who wrested their farms with so much labor and hardship out of this thick virgin forest, were forced perpetually to fight against the encroachment of the forest, and out of this ancient fear there still remains on many an Ohio farm a mania for cutting down every tree or shrub and all the vegetation along the roadsides and fencerows. It is very fortunate that the destroyers fight a perpetually losing battle in this rich country. One of the finest of American novels, written by Conrad Richter, is *The Trees* which is concerned largely with the struggle of one family of pioneers in the Ohio country to clear the luxuriant forest, and the battle to prevent the forest from constantly moving in upon their little island of cleared land.[1]

Even large areas of the forest at Malabar had suffered before we came there for they had been pastured heavily by cattle and sheep, which found little there to eat and so devoured the very wildflowers and forest seedlings in order to fill their hungry stomachs.

Except for the Anson Place and a little wild bottom pastureland and the impenetrable jungle, the whole landscape had the look of a devastated area. But the natural frame was there, waiting. Perhaps it took some

---

[1]There never was any pine forest in the state of Ohio. Because of the ample and well-distributed rainfall and the natural mineral richness of the soil, the forest was all hardwood save for remnants on north slopes here and there of the beautiful preglacial hemlock. Even today where pine plantations have been used to reforest abandoned hill farms, the hardwoods frequently seed themselves in and eventually crowd out the pines.

imagination to see what might be created there, but money, thought, imagination, and hard work were necessary. I think I saw quite clearly the picture that might eventually fill the frame, but it has taken nearly fifteen years to arrive at something approaching the realization of that picture. Today the whole valley is unrecognizable as the same valley to which I returned one snowy day sixteen years ago after being away from home for nearly thirty years.

I have seen and known landscapes in many parts of the world, marvelously beautiful landscapes created out of the climate, soils, and traditions of given areas. I am not a lover of arid desert countries or landscapes. While it is possible to recognize the spectacular beauty of rock and sky which sometimes occurs in such countries, I have always found myself miserable and unhappy in them after a short time, perhaps because I belong by nature, inclination, and perhaps even through atavistic forces, to rich green country with rainfall and four marked and distinct seasons. I am happiest in the landscapes and the countrysides of England, France, Austria, and in general the temperate zones of Europe and North America. That is where I belong—a feeling which most readers, and especially gardeners and farmers, will recognize at once, whatever their preference. There are of course individuals who are insensible to these things and are perhaps happy anywhere so long as they are prosperous.

Again, while I can recognize the wild grandeur and beauty of the Alps or the Rocky Mountains, I do not want to live among them. I like mountains as a tourist and for skiing and fishing purposes, but I would certainly be miserably unhappy either on a summit with an immense and overpowering view or in the bottom of a narrow dark valley where the sun rises two hours later and sets two hours earlier than in the world outside. High wild mountains can be a prison to some people as they are to me.

I think the mountains I have loved best were the biggest, most overpowering mountains in the world, the Himalayas, and this was not because I ever visited them or really knew them, ever attempted to scale their immense heights, or even to approach them, but because I saw them always from a great distance, a whole white and glistening world rising against the horizon, as the background to the lush, green jungly country—the rice paddies and ponds, the leopards and the wild elephants of

Cooch Behar in the north of Bengal. On the hottest, muggiest day the sight of this immense and distant frozen world filled one with a sense of coolness. Fortunately I was far removed from their ice and snow, their storms and wind, their barren rocks.

And so at Malabar in those first struggling days, I saw in my imagination a countryside that was soft and verdant with a little river running through it, where the fields were neat and green and rich, and the forest and marshes as wild as it was possible to make them. I did not see a landscape with harsh wire fences and clean fencerows and straight rows of corn, good or bad. I saw the kind of landscape that one finds in well-managed and well-farmed areas of northern Europe and this country—the kind of landscape one finds most naturally in those areas where the great glaciers have deposited whole mountains of gravel and silt that have an eternal and indestructible fertility. In such areas ... one finds them in northern Europe, in upper New York State, northeastern Ohio, Wisconsin and Illinois ... the hills are not eroded mountains; they are more often simply great heaps of fertility, where side by side there are rich valleys and slopes, well cultivated and fertile, together with forests and marshes that are completely wild. It is a kind of country in which there are always birds and abundant wildlife, streams and lakes and ponds and above all, countless springs of clear, cool water. That is the kind of country in which we set up Malabar Farm and went to work to repair the damage and devastation created by generations of bad and careless agriculture.

But one does not repair a whole landscape, heal the wounds, and restore the natural beauty and fertility overnight. It takes work and thought, imagination and, I think, above all else, time, patience, and love. It is essentially a creative job, one of the most difficult and fascinating in the world. One can neither hurry nor short-cut nature.

I think it is imagination and a kind of vision which drives the good farmer and the good gardener to work all hours of the day in every kind of weather, often enough beyond his strength. Arthritic old ladies forget the agonies of their rheumatism in order to tend some tender plant or shrub; the old and tired farmer will work into the darkness far beyond his strength to keep his fields neat, productive, and in order. This urge and drive has made of me, as it has made of many a gardener, a kind of contortionist, this drive which forces one to stretch muscles and sinews into impossible positions, merely for the sake of a seedling or the pruning of a plant surrounded by other plants which must not be trampled or injured. Certainly it has done much to keep a figure which would have become large and heavy and rotund by nature in ordinary city life, and maintained muscles and a suppleness which are quite as good as they were at eighteen.

In his imagination every good gardener and farmer sees the harvest with the planting of the first seed. As the earth covers it he sees the flowers, the rich grain, the beautiful grass that will soon be born from the seed. With each hoeing, with each weeding, with each cultivation he is driven by the vision. And the true gardener and farmer suffers with his plants and crops; if it is too hot and dry, he becomes ill and really suffers himself; if the plants are ill, he feels their illness in his bones. It is this vision and drive which keeps the good farmer forever planting, forever hoping, persistent and undefeated in the face of flood, of drought, of every disaster.

And so for fifteen years I have worked and suffered and sometimes spent money which I should not have spent, not merely upon restoring land and achieving rich crops, but in the creation of something more than that ... a whole farm, a whole landscape, in which I could live in peace and with pride and which I could share with others to whom it would bring pleasure. The most satisfactory things about such a vision and such a goal are that one must work perpetually with nature and that the task is never really finished nor the vision ever really achieved. There is always something more to be done; and so I shall be well occupied until I die at last, I hope in the midst of that very landscape and garden I have helped to create. It is not a task or a vision with which one can grow bored, for one is living with the whole of the universe which, as all will agree, is fairly inexhaustible during the short span of our lives.

Because I have lived long in France and loved it just as much and in some ways perhaps even more than my own Pleasant Valley, the first vision I had was a very French one ... a landscape with tall poplars and many willows, a misty, feathery landscape with great trees like those in the pictures of Claude Lorrain. I didn't think this all out clearly in any definite terms; it was just there in the back of my mind.

After fifteen years, the landscape is partly at least achieved. The fields are green and well ordered. In place of bare wire fences there are the hedges of multiflora rose, impenetrable alike to hunter and to old sows.

The willows are everywhere along the streams and around the ponds. Even the white barns and silos with their green roofs fit into the picture now, for the areas around them have been cleaned and tended and the trees planted long ago have grown to great heights in our rich soil and now the half-hidden houses seem surrounded and protected by them against the hot sun of August and the cold blasts of the winter blizzards which sweep from the north down the valley. Indeed, the houses, instead of appearing simply stuck into a bare landscape, appear to have grown there like the trees themselves.

It is a landscape which has grown more rapidly than I ever expected, partly because of the richness of our soils and partly because of small things done here and there … a mulch of stones or manures around a tree or a shrub, the thrusting into the wet spring earth of willow cuttings carried under the arm during long Sunday walks along the streams and ponds, the good manure coming from the big cattle barns, the legumes and the green manures grown in the field, the operations of modern machinery and deep tillage, together with the love and care and imagination of children and grown-ups. All of these things have helped Nature herself on the way; no force on earth will give such a response and such a reward as Nature when you understand her and work with her.

Mount Jeez, which for two generations had been overgrown with brambles and poverty grass and was known throughout the township as Poverty Knob, is a green hill today where the cattle in summer, as if by arrangement, come out of the shade as the hot sun begins to descend and move across the green pastures in the fading blue light. Mount Jeez is more than a mile away across the valley from the Big House and one can watch the cattle in the evening from the terrace or the verandas, a sight which warms you and makes you feel good. And the forest land is no longer pastured and scraggly and gullied, but grows wild and thick with its carpets of wildflowers and wild ferns, where in early spring the delicious morels grow here and there in clumps amid the wild geranium, the trilliums, the dogtooth violets.

There are also those areas near the houses which have become in time gardens that have been built partly by care and love and partly by the response of a kindly benevolent Nature. Somehow, it seems that Nature understands at times what it is you want to do and helps you along the way. With us she has certainly done so with the hundreds of thousands of jonquils and daffodils which now grow wild in the bluegrass pastures and along the edges of marsh and forest. A few bulbs placed in the ground, and presently there comes in spring a whole drift of blossoms where before there was nothing. In the forests of Chantilly and Ermenonville and Compiègne, the whole earth between the great oaks and beeches is carpeted in spring with wild daffodils, bluebells, and lilies of the valley ... the *muguet* which brings luck, and when it appears on the flower stands of Paris means that summer is at hand. All my life I have wanted so many jonquils and narcissus growing wild that one could pick baskets of them and never notice the loss. We can do that now at Malabar in the orchards and green pastures. They are everywhere on all sides with the cows and calves moving among them, feeding on the young spring grass without ever touching the blossoms.

The first gardens we laid out around the big houses were, consciously or unconsciously, conceived in imitation of the gardens on the small place where we lived so long in France. The house there, described in detail in *Pleasant Valley*, was an old *presbytère*, built in the seventeenth century and inhabited by the priest who tended the old twelfth-century chapel until it became half-ruined and disaffected. Like the fields of Malabar, the place had been half-abandoned and neglected for years before we came to Senlis. As the gardens went, there was nothing but a semicircle of ancient shaggy lilacs; the rest was weeds. But again the natural frame was a beautiful one. An ancient gray wall separated the *presbytère* from the ancient chapel and the graveyard that lay beyond ... a graveyard dating from the Middle Ages, which some thrifty Frenchman had changed into a vegetable garden perhaps a century or two earlier. Beside the chapel and the house ran a small clear canal with the charming name of La Nonette or The Little Nun. On the opposite side of the small canal lay an old orchard. It was actually below the level of the running water, so that one descended steps from the canal down into the lower garden. Save for a few ancient apple and pear trees there was no vegetation worth consideration. But the important thing was that we had a kind of vision of how the whole place should look and after ten years it looked much as we saw it in the creator's vision, and it became known as a famous small garden which was visited year by year by gardeners and architects from all Europe and America. Being small, not more than two or three acres in all, it became literally crowded with flowers, with roses spilling over the walls down to the water

of the little canal, lilies which grew magnificently, pinks, seven-foot delphiniums, really giant foxgloves, and the fuchsias, the geraniums, and the
ageratum that grew in the pots on the terrace.

It was essentially an English garden, and its exuberance frequently
astonished the French, who, with their logical and civilized point of view,
regarded most gardens not as a part of nature but as merely geometric
designs which were an extension of the ordered terrace and the classical
façade of the house. Yet it was at the same time a kind of international
garden, for the ideas came from everywhere. The pots of flowers on the
hot stone-flagged terrace were in imitation of the beautiful potted gardens I had seen in Mysore in Southern India, where everything was grown
in pots, and flowers were brought in during their period of blooming and
replaced by other plants in bloom when they began to fade. It was a small,
intimate, lush, well-kept garden ... a garden which I soon discovered
could *not* be reproduced at Malabar.

There were many reasons why this was so. The landscape itself was
much larger and wilder than the surrounding landscape in France. There
were no ancient gray walls against which the clematis and roses could display their deep, rich, and delicate colors. I myself, now the proprietor of a
hard-working big farm that desperately needed restoration and with far
greater responsibilities to community and state than I had ever known in
France, simply had not the time to tend it carefully and meticulously, to
keep an eye on every plant. Labor was scarce and expensive and badly
trained. But most of all I think the thing which proved defeating was the
exuberance of growth and the wildness that was characteristic of our
Ohio country, and which in my long absence I had nearly forgotten.

Plants which took naturally to our Ohio soils grew and increased with
an abundance and exuberance which was almost nightmarish. The day
lilies which I had tended so carefully in the French garden to keep alive
and in modest health became in Ohio real monsters of reproduction and
vigor. It was necessary to separate them at least every three years. If they
were the type which grew by underground shoots, they quickly overtook
and smothered everything that grew near them. On the roadside, the
common red variety grew everywhere like the most vigorous weeds.
Actually, the perennial asters, varieties which I had seen tended carefully
in the gardens of my English friends, grew wild everywhere along the
roads, in the marshes, and on the borders of the forest. And, of course, the

goldenrod, so carefully cultivated in Europe under the name of *solidago*, was our worst weed on the poorer land, where in October it produced literally seas of gold. And the dogwood and wild crab apples, regarded in Europe as rare ornamental trees difficult to force into bloom, invaded constantly our orchards and pastures wherever birds and other wild creatures dropped the seeds.

And here in this larger wilder country with no walls to protect the gardens, weed seeds blew in from everywhere and invaded everything. Some even grew vigorously beneath the snows all through the winter. Even in the beginning it became apparent that despite any desire on our part or any amount of nostalgia, nature simply did not want the kind of carefully tended and ordered garden that one had in Europe. This was no ancient ordered countryside with centuries of care and work and tending behind it; this Ohio country was still wild country, exuberant, vigorous, primitive, and only a few generations away from the forest wilderness.

In virtually the only time at Malabar when we attempted to force Nature, she simply refused to cooperate and turned hostile. For in the end we could not reproduce there the particular kind of garden and beauty we had created in France. Standard roses presently curled up and died, but the half-wild bramble roses, the wichurianas, the rugosas, the China roses, overtook and overran everything or turned into small trees of their own accord without the least help. The wonderful large-flowered climbing roses which had covered walls, houses, and bridges in the garden in France could not take the violence of the American winter, but the floribundas turned into giant flower-covered shrubs. Because of the highly available lime content of our glacial subsoils, rhododendrons and even some forms of azaleas were impossible; once their roots penetrated deeply they turned sickly and presently they died. But since I never believe in trying to force plants and shrubs to grow under conditions in which they are unhappy, we presently abandoned all efforts to raise them and instead expended our knowledge and efforts upon the whole range of plants and shrubs, an enormous one, which flourished in our native soils. Every kind of weed invaded the gravel paths, which in France had been so easy to keep neat and clean. Even elm trees and maples and black walnuts seeded themselves behind our backs and threatened to turn the whole garden back into forest once more.

Then came the war and all thought of continuing the struggle was abandoned and Nature was permitted to have her way. For nearly five

years the garden went wild and in doing so produced what Nature wanted. Once we yielded to her, she rewarded us handsomely.

This was really forest country and wild country. It was a wild and natural garden that Nature meant us to have and by the Good Lord that is
what we ended up with; and I know now when all the trees and shrubs
and bulbs burst into flower everywhere around the Big House each spring
that Nature was quite right. There is for weeks a kind of glorious blaze of
color, visible nearly a mile away from the Pleasant Valley Road, and in the
gardens themselves there is a kind of misty, even mystical beauty, created
partly by the delicate spring colors and partly the misty green of the
weeping willows and the tall upward thrust of the Lombardy poplars;
partly by the pale drooping Chinese splendor of the wisteria which would
never bloom so long as it was carefully tended in an ordered garden, but
which, when neglected and turned wild, filled the hundred-foot height of
the Balm of Gilead trees with a misty cloud of white and purple flowers
hanging low on the reflecting water of the ponds.

In our country trees grow quickly into giants and shrubs into trees.
Forsythia and Cydonia grow fifteen to twenty feet high and layer and
spread themselves in all directions. Flowering Japanese crabs and cherries
quickly attain the height and the size of the great gnarled apple trees in
the ancient orchard behind the house where the land rises upward toward
the cliff like outcrops of pink sandstone, covered with ferns and wild
columbine, and tilted as if by design of Nature to make a background for
the drifts of crocuses and grape hyacinths, jonquils, and narcissus in the
bluegrass beneath the blossoming apple trees. Once we turned Nature
loose, and let survive what belonged there and pleased itself there, she
presented us with one of the loveliest of gardens, especially during the
exuberant Ohio spring.

Many people contributed to the change and increasing beauty of the
valley landscape. Orrie Zody, a neighboring farmer who helped during
the first summer with the cleaning up of the rubbish which had first to be
gotten out of the way; and Dan Quinn, who, with his lusty crew of young
helpers moved to Florida in winter and back to Ohio in summer. They
built walls and did the grading of the rough ground around the house
and made the first plantings of flowering shrubs and trees. And Charlie
Martin, who, single-handed during the war, kept the farm in vegetables
and battled against the encroaching weeds and forest. And there were
countless friends, come for a weekend visit or to spend a week or two,
who set to work with good heart, to weed, to hoe, to prune.

During all the war, it was a scattered fight, intermittent and fought with guerrilla troops, volunteers, and drummer boys; of an intensity which only an Ohioan can perhaps appreciate and understand. But the significant fact was that we never gave in, and when the war was over we had no victory but only a breathing spell. War returned once more and everything on the farm and in the garden began to slump back into a tangle.

There is such a thing as a wild and natural garden becoming *too* wild and *too* undisciplined, and with the coming of the Korean mess and of a second war, that was what began to happen to the gardens surrounding the Big House. Again the young men were all drafted for the ugly, uncivilized, meaningless business of war and the older ones were drawn back into an industry to which war, carried on with the blood and suffering of the younger men, brought new prosperity and new high wages.

It was about this time that two young Englishmen stepped in and took over. David Rimmer, trained at the North Wales College of Horticulture, came first. He came not primarily as a gardener, but to take over and carry on the work of experimentation and research in the soil plots, a position for which he was well trained. But he came from Cheshire and was therefore English of the English. He had grown up in the ordered beauties of English countryside and farms. Their beauty was actually in his blood and he saw, almost at once, that while Nature at Malabar had helped us out by cooperation, she had begun, with the ever increasing fertility and preservation of rainfall, to overdo the job. He had what I had never really acquired as a gardener, the sense of a garden being a part of the whole landscape, of seeing a garden or landscape as a whole and not in small restricted bits and pieces, concentrated, lush, and isolated ... a conception with which I had perhaps become corrupted during years of life in France, where everything is ordered, pruned, staked, and grown against walls.

It is out of the broad conception which David brought with him from Cheshire that the magnificent beauty of the English garden, farm, and landscape has been created throughout centuries of thought, taste, practical good sense, and even genius.

The French look upon a garden as an ornate extension of the architecture of the house, or as a neat and ordered, practical, utilitarian affair which fills the larder or makes a living; but with the Englishman, from the owner of the semidetached villa along the tracks of the Midland Railway to the owner of the great English country house where once was practiced the most civilized living ever known to mankind, the garden must be a thing of romantic beauty as well as part of the landscape. Indeed, in some great English country houses in the past the whole of the landscape for miles has been treated as a single vast garden. It was David who brought this conception to Malabar.

Even as a resident for years of Europe, I still remained an American when it came to gardening and even, perhaps, to farming; I was and am impatient for quick results. I planted shrubs and even flowers too closely together, in the hope that in a few weeks or a few years, the whole place would be lush and overgrown and have a look of age. It is this impatience, this over-planting, this passion for speed, even in Nature, which makes many a new householder the victim of a swindling commercial landscape gardener.

From the very beginning I had definitely over-planted everywhere and when the war came and the gardens and shrubbery had to be neglected because there was no help, they became a veritable jungle through which it was difficult to fight one's way. It was David who ruthlessly saw the picture not in parts but as a whole, as a garden, or a farm or a valley or landscape should be seen. He had a fearful struggle to realize his conception, for each clump of lilies, each shrub, each tree was precious to me. But David, like myself, is a Capricorn, of a tribe characterized by stubbornness, persistence, and undefeatability, and he kept persisting, pruning a shrub or tree here and there when I had my back turned, refencing the shaggy ill-kept orchard so that the cattle were admitted to keep it all clean and well grassed and to wander across the scene just outside the windows of the Big House. Each discovery of his perfidy was at first an anguish and it might have taken years for him to accomplish by persistence and guile the fullness of his purpose. Essentially he was an artist creating a picture on the grand style. I was a miniature painter and pretty messy about it at the same time. As a practitioner of the green thumb category I would yield to no one; but in laying out the gardens, I was fussy. In general I practiced the very antithesis of the excellent maxim given me a generation ago by an old French gardener who said, "Never make a small and narrow path in a *small* garden or, in fact, in any garden."

The triumph and the achievement of David's purpose came when I acquired a big farm in Brazil and went there for part of each winter. On my return, at a time when in the North the winter was still with us and the branches of the trees were still bare, I was appalled, for the landscape all about the house seemed stripped and almost bare. It was only when the shrubs and trees began to burgeon that I began to see what he had seen and I had not seen. The over-planting, the crowding was gone. I even discovered trees and shrubs whose existence I had forgotten and which now began to show their full glory. One part of the gardens was turned into a calf lot planted with fruit trees, and here drifts and clouds of crocuses, grape hyacinths, jonquils, and narcissus began to appear, as they did in the orchards and in the bluegrass pastures. The cows did them no injury but seemed to walk almost daintily in order to avoid crushing them.

Around the ponds, where the trees and shrubs had become a tangled jungle, trees and flowering shrubs emerged in a singleness of form and color which was reflected in the water; and I understood what David was doing and had already done, perhaps without even knowing it. He was creating not a pinched garden or a crowded garden or a tangled, ragged, untidy field or farm, but a kind of Constable landscape in which the flowers and shrubs and ornamental trees merged and blended into the native landscape of our part of Ohio with its magnificent oaks, beeches, and maples. The hedges of multiflora rose, so practical and useful, became a part of the whole picture. The farm was no longer crowding the overgrown gardens. The whole of the farm was itself becoming a garden, planned skillfully and with the taste and understanding of an artist … an artist who could perhaps only be an Englishman with the misty dreaminess and perfection of the English landscape in his very blood. One could not really discern where the gardens ended and the farm began. The very pastures and orchards gradually became gardens.

In the exuberance and fertility of our country, the trees, the hedges, the wild flowers, even the lilies tend to get out of hand. Each hot, damp summer they make a prodigious growth, and each winter season they must be subdued and brought back again to the discipline and order without destroying romantic naturalness. There is a common assumption that the formal garden with its ordered rows and beds is difficult to create and maintain and that a romantic informal garden takes care of itself. No assumption could be less true. The formal garden requires merely architecture and some transplanting, hoeing, and weeding; the romantic, natural garden is an unending burden and requires a discipline and a concentration which no formal gardener ever attains or understands. A twig or branch must be snipped here and there, a shrub removed or transplanted. The drifts of daffodils and the apparently spontaneous beauty along the borders of a pond must be watched, guarded, and disciplined constantly. The burgeoning clumps of lilies must be separated and transplanted. A tree or shrub must be given air and space to emphasize its beauty. Like a painter, one must study the very landscape, regarding it with squinted eye; one must consider it in differing lights. But most of all one must never leave a landscape or any part of it to itself unless it is the thick forest or marsh, and even here there is work to be done with the planting of marsh marigolds or carpets of the wild violets which grow so easily in our country, or with the apparently careless growth of dogwood and wild crab apple.

The work is never ended, and if one neglects it for a single season, the picture becomes altered and changed and a certain ragged, untidy effect of carelessness creeps in. Just as agriculture is the most difficult and complex of all professions, so is the creation and maintenance of a beautiful landscape one of the greatest and most difficult of arts. I am not speaking now of the vast and lurid "views" of monotonous forests or craggy snow-capped mountains, which only emphasize the puny insignificance of man, but of the beautiful landscape which provides evidence of his civilization and his sense of industry, intelligence, taste, and art. Without these things, the insignificance of man and his unimportance become simply overwhelming.

A little later Patrick Nutt, who had his training in the greatest of all schools, Kew Gardens, came over to join David. Patrick's primary job was that of hybridizing, propagation, and greenhouse work, but during the summer months he too is occupied with the experimental and research plots and the vegetable gardens where much of the experimentation is done. He too came to contribute his share of the genius which English gardeners, both amateur and professional, possess, and he too has left the imprint of his knowledge and taste across the face of the Malabar landscape. No one, I think, understands better what it is that a growing plant likes and needs. Whatever he grows is essentially a specimen plant, and out of a very small greenhouse and a few hotbeds he produces quantities of flowers through the winter, apparently with little effort and time. He is like a good dairyman with his cows; the plants seem to like him and he seems to like the plants and out of the mutual affection, miracles are born. I doubt that any two young men now working in this country are contributing more to the knowledge and science of soils and production than these two young Englishmen. Fortunately, they have come to the States to stay, and David is already a citizen of this country, for which I think American agriculture and horticulture may well be grateful.

Once I gave David and Patrick a free hand, I began to realize that if I still wanted my lush small perfect garden, it would have to be *not* among the ravines and hills and rocks which surrounded the Big House, but on the low flat land below the pond at the Bailey Place, and so there came into being there a garden where, as in the French *potager*, flowers and vegetables and fruits are grown all together under the most concentrated and intensified conditions. And it is there that I find myself most of the time, where the delphinium brushes leaves with the rhubarb and the lilies and roses flourish side by side with the asparagus and the strawberries. And in this flat rich garden where much of the experimentation is carried on, there is always a wealth of magnificent, even specimen flowers, for cutting, for the roadside stand, and for the pleasure of the thousands who visit Malabar each year. With the passing of each season this garden becomes richer and more beautiful with that special beauty that is found in the French *potager* where flowers and vegetables grow side by side and all of them are of specimen quality; where the great leaves of the rhubarb or the ornate foliage of the eggplant or the Corinthian leaves and purple blossoms of the artichoke are quite as beautiful as any of the flowers.

I think now we are on our way with the gardens and the landscape of Pleasant Valley. We have found the things which belong there (as I myself belong). We understand the climates and the soils and each year new mysteries are unveiled and wonderful things take place in the depth of the good earth from which all of us come and to which all of us must return. It is a happy life, and one in which there is an immense variety and to which there is no end, for even after death someone else will carry it along. Now when I step out of the door the whole of the valley seems to be a beautiful landscape by Constable or Claude Lorrain. It is as beautiful in winter as in full summer, as lovely in the pastel colors of spring as it is in the flaming colors of October. But it didn't just happen. It took a lot of thought and a lot of work. But it has been worth it. And the task is never finished. It goes on and on.

# MALABAR-DO-BRASIL

IT ALL BEGAN YEARS AGO IN THE SUMMER OF 1941, WHERE THERE APPEARED one morning under the big walnut tree on the lawn of the Big House two men and two women who were quite obviously not American farmers. They were well dressed, the women with a great sense of style, and obviously they came out of the great world. This was evident from the way in which they approached, the fashion in which they introduced themselves in accented English, with great simplicity and directness. It is a curious fact that the simplest farmer and gardener and the most intelligent, experienced, and sophisticated of people have the same good and easy manners, because both are likely to have a proper sense of values and the realities of human relationships. It is only the foolish or the pretentious or the vulgar who adopt elaborate styles of introduction, coyness and phrases and manners calculated to impress you with their importance. They are like the insignificant and pretentious people who use the long-distance telephone constantly in order to make an impression of being important big shots, when a telegram, postcard, or a letter would serve quite as well.

It was evident from the first that these people knew their way around. They introduced themselves as Senhor and Senhora Manöel Carlos Aranha, and Senhor and Senhora Francisco Almeida and explained that they had heard of Malabar Farm and had the temerity to stop on their way through from Chicago to New York to have a look at it. They apologized for interrupting what I was doing. They came into the house and we had drinks, and before very long I knew that I wanted to show them around the place. They were intelligent. They were interested. They had a great deal of knowledge concerning agriculture and economics. What I did not know was that this meeting was the beginning of a great friendship and a relationship which was to bring great changes and great new interests into my life and the life of my whole family and even of all the people at Malabar. It was to open up for me a whole new world—the world of Latin America, and in particular of Brazil. The two couples were Brazilians from São Paulo, and Carlos Manöel Arañha, although he was one of the important industrialists of Brazil, had a deep and intelligent interest in agriculture, a great love for his own country, and deep concern for its future.

Looking back, it seems to me now that Malabar had very little to show at that time, although at the moment the work and the achievements *seemed* stupendous to all of us on the farm. We had checked the worst of the erosion. We had laid out strip crops and changed the old square fields, where it had been the habit to plow up and down the bare hills, onto the contour. We had seeded down many of the worst fields with a healing blanket of grass and legumes. We had set up a dairy and bought some livestock. Beyond that we had nothing much to show. We had not even acquired two of the ruined farms which later became a part of Malabar. But there were great plans ahead and there was enthusiasm on the part of all of us concerned, and even the few changes we had made created a difference between our land and much of the eroded, barren, weed-grown land in the immediate neighborhood.

Like myself, Manöel Carlos was and is a great talker, speaking rapidly and quickly with enthusiasm, as if his tongue could not keep pace with the rush of his ideas. As we went over the fields and remarked on the changes, I learned many things ... that he was an industrialist in the glass business, that he had connection with great American glass manufacturers, and that he had recently acquired some three thousand acres of eroded and abandoned land in Brazil, between São Paulo and Campinas, and was about to begin operations designed to restore it to productivity. The place was an ancient half-ruined coffee *fazenda* called Rio de Prata, and its history paralleled that of much of the land in this country. It had been badly and greedily farmed on the belief that it owed the owners a living, and at last it had been abandoned save for a little bottom land subject to floods but not erosion. The Brazilian coffee growers had moved on west to new land and when that was worn out moved still further west. This was an old pattern, the pattern of all new countries which still possess frontiers of new rich virgin land. There were millions of acres of such land in south central Brazil, and there were a great many intelligent Brazilians who were concerned about the enormous damage and waste. Manöel Carlos Arañha was one of them.

One of the pleasantest experiences of life at Malabar is the variety of the people who drop in from almost every part of the world. There were Robbie and Mary Robertson, who from South Africa came unannounced to the door, were invited for lunch, stayed overnight, and eventually left us with regret on both sides, after three weeks. And three or four young farmers from New Zealand and Australia who have happened along and spent days and even weeks with us, and a great number of Englishmen, Scots, Frenchmen, Latin Americans, and Swedes and even a Gujarati from India who arrived with all his luggage, prepared to stay with us until his family arrived from India some months later. (We did not follow out his plan.) Out of all these apparently casual contacts have grown great friendships and new interests, not only for myself but for all the household and for all my children, who were partly educated in Europe and enjoy friendly and intelligent people everywhere regardless of race, color, or creed.

Before Malabar Farm came into existence about half of my life had been spent in Europe and the Far East, but I knew little or nothing of Latin America or of Brazil. I simply shared the appalling ignorance of most of my fellow Americans about the rest of the fabulous Western Hemisphere, even those with the sometimes dubious honor of a university diploma. The visit of Manöel Carlos broke the ice and marked the beginning of a new knowledge and interest in the magnificent parts of the Western Hemisphere which are wholly unknown to most Americans, a hemisphere of incredible and only partially exploited natural wealth and fertility, of Spanish and Portuguese cultures and magnificent architecture, of friendly peoples and among them the friendliest people I have ever known, the Brazilians. Their friendliness resembles that of most Indians before our own government and State Department lost us the friendship of Nehru and Gandhi's people, but they are, as a people, less intricate, subtle, devious, and changeable than the Hindu Indian. As a people the Brazilians are simple and direct and polite, good-tempered and full of an almost childlike goodwill, perhaps because, like Americans, they are a people with many strains of blood, of many cultures, and many basic nationalities, and because they live in what is potentially one of the richest countries in the world and certainly one of the most beautiful. They have remained friendly to us even when our government and State Department have neglected and been rude to them in favor of other ventures elsewhere in the world which have proven all too often merely disastrous.

From that first visit onward, Manöel Carlos came nearly every year to Malabar to take away with him books and pamphlets on modern American agriculture, on soil and water conservation and forestry, on chemistry and organic material, on livestock breeding, and to compare notes with us on modern agriculture and soil and water conservation. At the *fazenda* called Rio de Prata, far away in Brazil, great things began to happen, and slowly the abandoned, eroded land began to turn into what is today perhaps the finest and best managed farm in all Latin America. In the meanwhile I discovered that among other accomplishments Manöel Carlos was a great athlete. He had been perhaps the most famous and beloved of all the amateur soccer players in Brazil, where soccer arouses the same wild enthusiasm and partisanship that baseball arouses in this country. He had been a member of the Davis Cup team in tennis and played in international matches many times. Unlike some of his fellow countrymen he was and is a man of tremendous and unflagging energy and drive, and in addition he possesses a great and practical intelligence.

Before very long he became known to all of us at Ohio Malabar merely as Carlito, the affectionate diminutive by which he was known to all Brazilian soccer enthusiasts. He has since become the godfather of one of my grandchildren who lives in Brazil and at the time of writing his son Rubens is at Malabar in Ohio learning as much as possible about practical farming in the U.S., all the way from weeding in the truck gardens through hauling manure to operating the heavy machinery.

In the years following that first visit, dozens of other Latin Americans and especially Brazilians turned up at Malabar and presently there began to come in requests for us to set up a Malabar Farm in Brazil. But I had no desire to begin such an undertaking in a strange country where I did not know the language and where the politics and economics could at times be very difficult. I made two visits to Brazil, taking with me on each occasion some thirty-five American farmers, ranchers, bankers, and livestock men on a sort of goodwill tour arranged by the Braniff Airlines. By then I already had many friends in Brazil and during these trips I made many more, and the fellow countrymen who went with me loved Brazil, all save two or three Texans who were homesick and in mind and spirit never really left Texas at all. The urgent suggestions to set up a Malabar operation in Brazil kept being pressed, but for many years I remained reluctant.

And then all at once, largely through the enthusiasm of Carlito, everything began to fall into place. We had gone to lunch at the *fazenda* of Julio Mesquite, the owner and publisher of a great newspaper, *Estado São Paulo*, a citizen of the world, and a man and patriot of great distinction. The immense house of the old *fazenda*, built more than two hundred years earlier, lay among the hills between Campinas and Jundiai—a huge house built in the Portuguese baroque style with a great courtyard, filled with jibuticaba trees hung with orchids, ferns and cages, filled with brilliant birds, which were mostly quite tame and had their liberty. It was the day after carnival and at lunch in the big room which served both as ballroom and as dining room, we sat down to a wonderful lunch together with thirty-four other people of all ages from ninety down to children of five or six years old. They were all relatives and at carnival time each year, they gathered in the beautiful old house that has belonged to the family since the country surrounding it had been a jungle frontier wilderness. An avenue of poinciana trees led up to it from the road and all about it grew a collection of trees, shrubs, and flowers brought from every part of the tropical world. During lunch the conversation was sometimes interrupted for a second or two by the loud cries of the brilliant macaws which Senhor Mesquite had collected from Mato Grosso, Amazonas, Bahia, and indeed every part of Brazil. These brilliant birds lived in the great trees outside the big windows and now and then darted past like bursts of a lightning made of brilliant reds, yellows, blues, and greens.

After lunch, as the young people, still a little weary from the carnival festivities which had gone on all night in Campinas, drifted away to the swimming pool or the nearest cool spot for a nap, Carlito, the Energetic, said, "There is a farm over the mountains I would like you to see. It belongs to a very interesting and brilliant doctor friend of mine. His health is not so good as it once was and he is finding the place something of a burden." I don't know whether, at that moment, Carlito actually had Malabar-do-Brasil in mind, but again, without my knowing it, there lay ahead of me one of those experiences, sudden and sometimes unexpected, which can change the whole course of a life and bring it new vitality and new richness of experience.

We set out in Carlito's car, not along the back highways, which themselves are none too good in Brazil, but up over the mountains along a wild trail which at times seemed suited only to goats. Like myself, Carlito's interest in automobiles is confined entirely to the kind of car which will get you there, regardless of how new or old it is or what its actual appearance. Like myself, he has no inclination to buy a new car each year in order not to lose face. The whole importance is fixed upon its sturdiness. There are certain vintage cars which serve one well and to which one becomes attached and from which one will not be separated. Such a car has nothing to do with "keeping up with the Joneses" and is the bane of the automobile dealer's existence. He will talk eloquently and at great length upon how much it is costing you to keep the old car and what savings you would make if you would turn it in for a new one. But when I have discovered such a workhorse car, I find that both his reasoning and his economics are not only faulty; they are sheer humbuggery. I have had two or three such workhorse cars in my life. One was a Fiat, one a Ford, and at the moment I have a Pontiac from which nothing can separate me.

Carlito's car was a Studebaker, and a true workhorse built high enough to clear the rocks and bumps, able to weather alike the foot of mud or the foot of dust which seem to be the only two states known to Brazilian roads which are not paved. Another Studebaker piloted by Santo Lunardelli once brought me from the far reaches of São Paulo State, almost to the borders of Mato Grosso, through valleys of mud and hills of dust to São Paulo. We arrived at four in the afternoon and the same car, washed and polished, took us out to dinner the same night. That is my kind of car.

As we followed the mountain trails among half-lost *fazendas*, through herds of half-wild Indian cattle, through the shallow swift-running waters of the spring streams, it became apparent that Carlito's automobile could take anything. It was the end of summer in Brazil and the jacaranda and piñieiro trees were in bloom on the hillsides and all around the pink, lemon yellow, and white farmhouses. Over each hill lay a new small valley and a new world and presently we came through the clean little white town of Itatiba, on the side of the hills fringed with bamboo, and native trees hung with orchids above a beautiful river called the Atabaia.

Carlito said, "It is only a little farther on," and we drove off the main highway into a dark cool tunnel of a road that led through the trees. Presently we crossed the river on a high wooden bridge, turned a corner and came into a magnificent avenue of eucalyptus trees with mottled white trunks reaching up toward the brilliant blue Brazilian sky. On one side, between us and the river lay a great marsh overgrown with tangled vegetation and on the other a range of steep sloping hills only partly under cultivation.

Carlito said, "The Englishman planted the trees."

The avenue was not straight. It took a winding course which added to its beauty, for with each new turn, framed by the great white trunks, there was a new view of the valley. Then the avenue came to an end with a sharp turn into a second wider avenue that ran down to the river and the Big House of the *fazenda* came into view.

It was long, one-storied and white with a roof of heavy weather-stained red tiles, almost hidden by the growth of shrubbery and the pink flowering piñeiros and purple jacarandas. On one side there was a corral and a big one-storied cattle shed open on all sides, and on the other a long enormous white building erected long ago for the administration of the *fazenda* and bearing the wonderful Arab-Portuguese name of Almaxerifado. (For me it is an unpronounceable name which I cannot remember and it has come to be known merely as "The Asofoedite.") The whole place lay in the midst of a broad rich valley with the hills sloping upward toward low mountains on both sides.

As we drove into the courtyard, the details of the old house and the gardens became evident. At one side there was a small canal of clear, swift-flowing water which fed a small lake with an island overgrown with oleanders, hibiscus, palms, and all sorts of native flowering trees and shrubs, and behind the house, extending down to the river, there was a big orchard in which I was able to recognize all sorts of citrus fruits and avocado trees. Mango trees grew everywhere, and on the far side there were two more avenues of gigantic eucalyptus trees, one leading down to the river and one leading away into the remote, immense distances of the valley and the hills; avenues of mystery which excited the curiosity. And suddenly inside myself I had the same feeling I experienced at the first sight of Malabar in Ohio after a generation of absence from my home county. This was my kind of country. This was the place. I was at home.

The brother and nephew of Dr. Santos came out from the small jungle of garden to greet us and in the Brazilian fashion asked us to have tea. Then we went into the house.

It had no splendor of architecture like that of the great Portuguese baroque house where we had eaten lunch. It was a house which had just grown over a period of generations, perhaps even of two or three centuries, and it rambled on and on beneath the great trees where orchids hung from all the trunks and lower limbs. The house had the charm that touches all houses that seem to have grown out of the very earth, generation by generation. Outside it seemed more like a number of small houses or cottages built at various times and for various purposes, and at different times connected by hallways and new rooms. Around the whole of the house there was a wide veranda almost shut in by a hedge whose exuberant growth was out of hand.

Inside the house was very dark and cool, for all the windows were shadowed by shrubbery, flowering vines, and trees.

In the largest room, which had a curious inexplicable Anglo Saxon air, I made a very interesting discovery. There was a large library of English books, revealing the taste of some owner who was no longer there. Mostly they were by H. Rider Haggard, Sir Oliver Lodge, Mrs. Humphrey Ward, and other writers of the late Victorian period. When I commented upon the strangeness of such a library deep in the Brazilian countryside, Carlito said, "They belonged to the Englishman," and I learned that years earlier the place had been the property of an English planter and banker who raised polo ponies and hunters. What had become of him, whether he was alive or dead, no one seemed to know, but the news seemed to clear up many things … the great avenue of eucalyptus planted by a homesick Englishman, as if to recall the great avenues of elms or oaks leading up to the big country houses in England; the beds of ancient roses with trunks like trees, which in a semitropical climate with no real winter seemed bewildered and were flowering out of season, in a scraggly way; and the few rather tired trees and shrubs from a colder climate which were still living, but obviously unhappy in a climate in which there was no winter and they had no chance for sleep. They seemed pale and anemic here in Brazil and were constantly endangered by the exuberant growth of the native and imported tropical vegetation all about them.

As in so many other parts of the tropics in Africa, in India, in Malaya, a homesick Englishman had clearly attempted to recreate the damp, cool beauties of the English gardens and countryside in another vastly different world, only to fail, swamped and overwhelmed by the richness and exuberance of tropical flowers, vines, and trees. I had seen such pallid gardens almost everywhere in the world before the sun at last set forever over most of the British Empire.[1]

The "tea" turned out to be an enormous meal served by an ancient housekeeper who shuffled to and from a kitchen and pantry lost somewhere in the dark reaches of the enormous rambling house. There was a great bowl containing every sort of tropical fruit ... grapes and mangoes, oranges and avocados, kumquats, and even pears and apples. There were several sorts of pastries and cakes, for most Brazilians seem never able to get enough sweets. There was coffee and tea and milk and small sandwiches (perhaps a custom left behind by the Englishman). There were two kinds of pudding, one a delicious kind of custard stiffened with cornmeal and a the other wonderful sweet wholly new to me. Being considerable of a gourmet and something of a cook, I can usually divine or at least make a guess at the component parts of a new dish, but this was a dark mystery until the housekeeper explained it. I give you the recipe to try, for it was and is one of the most delicious of desserts, especially in a climate or a time when the weather is hot. It is also one of the most simple of dishes.

---

[1]I have recently discovered the true identity and a part of the history of "the Englishman" and at the same time discovered how greatly country gossip, hearsay, and tradition can distort the truth. Actually "the Englishman" was a Brazilian citizen, American by blood, a baker by profession, and one of the descendants of a whole migration of Americans from the Southern States who left their country at the time it was invaded by the scoundrelly carpetbaggers of the North. In São Paulo, beyond Campinas, they established a colony called Villa Americana, which has since grown into a flourishing town. Many of them and their descendants intermarried with the Brazilian families and a few in each generation returned to the States to find wives for themselves. Their presence in Brazil accounts for the fact that in many great Brazilian families today there exist names such as Washington and Lee and Jefferson. The former owner became known as "the Englishman" in the countryside simply because he spoke English. He was a great horse breeder and fancier.

It was made simply of fresh avocados crushed through a sieve into a thin puree, mixed with the juice of lemon or lime and sweetened with sugar. It should be served as cold as it can be made without freezing.

After the "tea" Carlito, David[2] and I, accompanied by the nephew of Dr. Santos set out in Carlito's indefatigable Studebaker to explore the whole of the great fazenda and as we progressed, my excitement and pleasure grew. One could read the record of the landscape. Once this had been a great coffee plantation, and here and there over the hills and fields were scattered the half-ruined houses, built like the Big House with stout walls made of wattle and the stiff red clay of the countryside, and covered each year or two with a coat of whitewash. These houses, when neglected, do not rot or fall down into pieces; they merely seem to melt away in an infinitely slow process back into the earth from which they came. There were many of them built a century or two before, to house workers and slaves in the days before the slaves were given their freedom.

The history of the countryside was not much different from that of our own cotton South. Poor agricultural methods and exploitation without intelligence had gradually ruined the land over generations and even centuries. The freeing of slave labor was given as the excuse for the final decline and ruin, but it was no more a genuine reason and it had no more validity than the same excuse in relation to the dying agricultural economy of our own South. Both areas would have gone down into ruin if the slave labor had never been set free. Both areas were ruined by their own proprietors.

But the ruins of the melting old houses had a melancholy beauty. Some were still habitable—for the repair of such houses is fairly simple. One simply takes the clay from the field in the wet summer season and fills in the holes and the cracks, giving the whole a fresh coat of whitewash. These houses seem actually to grow out of the earth and to return to it quite easily when abandoned. If neglected for a long time, vines and even trees and shrubs take root in crevices of the walls and even the tile roofs and hasten the return to earth. In a few of the houses in the remote parts of the farm where strangers never came, dark faces peered out of the windows and sometimes the dark eyes of whole troupes of children

---

[2]On this trip I was accompanied by David Rimmer, who came to Ohio Malabar from England some years ago to take charge of the research and experiments in soils and who, together with Patrick Nutt, another young Englishman, is now in charge of the flourishing market gardening department at Ohio Malabar. His presence in Brazil and his advice was requested by Brazilian friends and agriculturists.

gleamed out of the tropical shrubbery that surrounded and constantly threatened to overwhelm the houses.

The whole big *fazenda* was made up of a number of farms tied together in the past by the central administration located at the Big House. The most fascinating of them all was a remote abandoned farm called Barreiros that lay far up the side of the low mountains. To reach it nothing less than a jeep or Carlito's indefatigable warhorse was necessary, for what had once been a road had long ago become merely a goat track. The bridges across the small swift-running streams were rotten or had disappeared altogether, and the whole place lay completely hidden behind a forest of eucalyptus trees planted long ago by the Englishman.

An old forest of eucalyptus, close planted in regular rows as is the habit in Brazil, is indeed a forest out of a fairy tale by the Countess D'Aulnoy. It is the dead forest of the ballet, *"La Belle au Bois Dormant."* The trees grow straight up to an immense height and beneath them there is no vegetation at all. There are only the fallen dead leaves of the trees, decaying in a place where no sun and little light ever penetrates. Even in the violence of a Brazilian storm there is no wind or even a breeze. Within such forests there seems to be no animal life at all and one can only think of Keats' line, "and no birds sing." The eucalyptus trees have simply taken over that particular part of the earth. There is no competition. Nowhere save on the edges of the forest where some sunlight falls are there even any of the serpentine tropical vines which in the native Brazilian forest strangle sometimes even the giant trees. One has the impression of being in the bottom of a crypt, dark, gloomy, given over to the dead.

The eucalyptus is not native to Brazil, but it grows more luxuriantly and more rapidly there than anywhere in the world, and it has become a great economic asset in arresting the erosion of the soil, checking floods coming from the older land, and providing a valuable crop of timber. Certain varieties produce a valuable aromatic oil. It has been planted everywhere in south central Brazil on the tops of the hills and mountains and on abandoned, eroded land. It has strange and powerful restorative qualities and it is said that worn-out, abandoned land planted to eucalyptus for a generation will become rich once more, perhaps because, with the natural richness of the soil, the earth needs merely a blanket of vegetation and deep-growing roots to break up its tough locked-in fertility. In Brazil a plantation of eucalyptus will provide a crop of salable timber every six or seven years, and when cut, the new growth of sprouts provides a second crop in the same length of time, a third, and sometimes even a fourth. It is indeed a prodigious tree, but perhaps its greatest value is what appears to most growers to be a secondary one ... that of covering the soil and preventing erosion and the runoff of flood water that can be so disastrous in the rainy season. Even the most violent of tropical cloudbursts is simply swallowed up by a thick forest of eucalyptus, as it was swallowed up long ago by the native forest that once covered the whole of the vast Brazilian country.

The still and ghostly forest that separated Barreiros Farm from the rest of the *fazenda* presently took on a kind of pale yellow illumination as we neared the far side, and at last, with a sense almost of relief, without glancing behind us, we came out into a wild country half-abandoned and partly grown over with the trees which move in everywhere in Brazil the moment the land is left untended. Over the wild landscape wandered some half-wild Indian cattle, a small herd of ponies, and three or four of the small, delicate, intelligent, and humorous mules which belong on every Brazilian plantation, large or small.

The Indian cow, known in the States as a Brahma and in Brazil as a Zebu, is a formidable animal, large, rawboned, long-legged, and fleet of foot with large drooping ears. In her method of attack there are no Marquis of Queensberry rules. She will butt, trample, and even roll on you, and I have heard that she will, in a true fury, even bite you. Unlike our temperate zone cattle, she has a deep bass voice and her bellow is more menacing than that of any ferocious Jersey bull. These Indian cattle will also move together in an attack, swooping about a field with an effect curiously like the flight of a flock of birds.

One such animal took exception to the visit of intruders. Somewhere in the neighboring jungle she had a calf and it is just possible that she had not seen a human for weeks or for months. Very likely Carlito's warhorse was the first automobile she had ever seen. After a violent stamping of the earth and a series of hair-raising bellows, she took out after our car and followed us, bellowing and threatening, all the way to the ruined farmhouse of Barreiros.

The roof of the farmhouse had long since partly fallen in and the walls had begun to melt away. Where a part of the roof remained, it was evident that it was used as a shelter during wild storms and in the heat of the day by the wild cattle and ponies.

All about the farmhouse there grew flowers and vines of every sort, which had spread into the neighboring fields, from some garden abandoned long ago. Whoever had lived there last—and it turned out that no one in the countryside could remember who the last tenants were—had loved flowers and orchards, for at one side of the houses there was the ruin of a fine orchard of fruit trees. Oranges and lemons, sweet and sour; limes and citron; and even a few ancient and sickly peach and pear trees still remained, half-buried in thick vegetation. And there was a single tree bearing the finest mangoes I have ever tasted anywhere in the tropics—large, juicy, stringless, and wholly free of the faint taste of turpentine which some people find objectionable in mangoes. Strangest of all, there were on the very top of the hill springs and two large marshes alive with noisy, brilliantly colored birds.[3]

The enchanted quality of the dark eucalyptus forest seemed to extend even to the abandoned house and garden and to a spring which was unlike any other spring I have ever known. The water flowed from an ancient broken leaden pipe and was sweet and fresh to the taste, but instead of being cold, was lukewarm. The stream flowed clear and steady until one approached within fifty feet of it when the flow became interrupted and took on a pulsating quality as if some spirit sought to cut off its flow and frustrate the thirsty traveler who had dared to pass through the dark woods where "no birds sing." The story—from whence it comes I do not know—is that the water is radioactive. But even the warmish

[3]The Barreiros Farm underwent restoration in 1954 when it became occupied by a family of Brazilian farm workers. The process seemed a perfectly natural one, as natural as the rising and setting of the sun. One day a family arrived with a wagon drawn by two pretty mules and containing all their furniture and belongings. They moved into the half-ruined house, rebuilt the adobe walls, cleared the area around the house, put out a garden, and then set to work clearing the fields for grapes, and the marshland for growing rice and onions and celery. It has already been rescued from the encroaching forest and has begun again to be productive. High on the abandoned hilltop of this extraordinary country, this farm raised in 1955 bumper crops of corn, potatoes, and rice. The story is the same as that of many a farm in the U.S., "worn-out" and abandoned by its original owners, reclaimed and rehabilitated by a whole new generation of pioneers, whose task is not to clear away the thick forest or prairie and fight the aborigines, but to reclaim what had been ruined by their forefathers. This process of reclamation is, in Brazil, making great progress as it is in the South and Middle South and the fringe areas of the Middle West in the U.S.

Later on, this mango tree, together with the sweet lemons which were of especially fine quality, became the object of regular visits by my children and grandchildren and myself. The trip to the wild country of Barreiros through the enchanted, ghostly forest into the lost world beyond became a special pleasure and adventure. The children were always asking in various languages at the Big House, "When are we going up to Barreiros?" In some ways I almost regret the restoration and the coming to life of the ancient farm.

water tastes good on a hot day after the long rough trip over the ruined road from the Big House into the hills.

From the eminence of the abandoned orchard there was one of those tantalizing distant views which always makes hill country fascinating. Far below the Barreiros Farm there lay another valley with a great marsh filling the hollow. Beyond this was visible a quadrangle of small white houses built long ago to house the slave labor, and above it on the hillside one could see the tiled roof of a Great House. Nothing but the roof was visible, for the rest of the house was entirely concealed by the huge piñeiros which, in full blossom, seemed to engulf and almost overwhelm the house in a froth of pink blossoms.

Between the Barreiros Farm and the house drowned in blossoms, there was no road or pathway. The distant Great House was completely shut off by forest and marshland, so isolated that it might have been a thousand miles away, although as the crow flies, it was quite near at hand. Carlito, unfamiliar as myself with this whole new world that lay over the hills and forest above the rambling house by the river, did not know the name of the *fazenda* nor who owned it. Afterward, when my own children came to live by the river, I discovered that it was called the Fazenda of Donna Ottilia, and that the hidden Great House was one of the gayest and loveliest houses I have ever seen anywhere in the world, surrounded by a garden and orchard of such exuberance and beauty that it is difficult to describe.

I was also to discover that, like most Brazilian Great Houses, it was almost a village in itself, filled always with people of all ages; cousins, uncles, aunts, nieces, nephews, and grandchildren of the splendid, matriarchal Donna Ottilia. A great friendship was destined to spring up between the people at the Fazenda Ottilia and the people at Malabar-do-Brasil, and there were to be great visits and goings and comings from one *fazenda* to the other, which were in a sense real expeditions, because in Brazil when one goes visiting one takes all the family down to the youngest children and sometimes even the dogs, which, in Brazil, seem to be more amiable than elsewhere. But for the moment the Great House, with the distant clustered little white slave houses hidden among the great piñeiro trees, remained a tantalizing mystery, distant and wholly remote.

The spring at Barreiros is only one of many on the big farm. Not only is it the kind of green country which gives me great satisfaction and in which very clearly I belong, but there are springs and streams everywhere, running swiftly, cascading down the ravines and valleys toward the marshes and the River Atabaia ... springs and streams of the greatest value to a farm for irrigation during the dry season. On parts of the *fazenda*, on the hilltops and bordering marshes, there still remain portions of the original virgin Brazilian forest and jungle, which is unlike any other jungle in the world. Certainly no jungle is thicker or more impenetrable. The bastions of such a forest are the gigantic hardwood trees.[4] Climbing these and suspended from them is a matted tangle of vines, and when any light finds its way down, the forest floor is entirely overgrown with matted vegetation. Beneath the roof of the forest the light, the very air, is green. It is the world of W.L. Hudson; a mystical, unearthly world of mystery.

I have had the experience of flying over the endless swamps and forests of Mato Grosso in a single-motor plane, and it is the kind of flying I care for least in all the world. One puts one's trust wholly in the single motor and the quality of the gasoline. If either failed and the plane was forced down, there would be no chance of rescue, for there would be small chance of finding a lost plane. No plane could land in the forest and from the forest there is no escape, for one would have to chop one's way every inch of the way out.

The vegetation of Brazil, and especially of the forest, swamp, and marshland, is one of the heaviest and most varied in the world, and for a traveler or visitor it is a constant source of wonder and delight. Brazilians take for granted as a weed many a flower, flowering vine, or tree which elsewhere would be cherished as a miracle of beauty. The forests and marshes are very often no more than gigantic flower gardens. Botanists constantly find new plants and even species, and one could make a magnificent showy garden merely out of the wild plants and "weeds" that are passed over and scarcely noticed by many a Brazilian.

On this half-wild hill farm there were not only areas of original virgin forest but also considerable areas of untouched, perhaps unpenetrated swamp and marshland, with varied and beautiful flowers and brilliant

[4]This lovely tree known in Brazil as the piñeiro is really the kapok tree native to Java, which was brought to Brazil sometime in the remote past and now flourishes everywhere, even growing wild in the fields and at the edge of the forest. The individual blossoms are the size of a large Cattleya orchid, which they closely resemble in flamboyance and beauty, although they are more the shape of a lily blossom. The petals are pink, ranging from a deep pink at the edge to white at the center and are mottled with tiny green and brown spots. When they are in full bloom the earth beneath the trees, which grow to a great size, seems covered with a drift of pale pink snow.

birds. These marshes were frequently located high up in hollows and depressions of the hills, and as Carlito and David and I drove around the place, it was evident that in a modern and practical program of farming, such areas could easily be converted into ponds and small lakes to be used as reservoirs for irrigation during the dry winter season.

The season was the end of summer when the Brazilian sky is at the most beautiful. It is a very special sky, which from the tops of the hills appears larger than the sky elsewhere in the world. It is a special blue in color, and of great brilliance, which becomes bluer as the sun sets and night comes on, until presently the blue fades into darkness and the stars come out like diamonds on velvet. In the daytime and in particular in the hot afternoons, enormous white thunderhead clouds drift across this blue sky. They resemble immense heaps of whipped cream and in summer one can always sea the warm rain from one or another of them streaming down over some remote small area of farm and forest. Flying in Brazil during the summer season is largely a matter of dodging these great thunderheads and the turbulence which surrounds them.

Toward evening we turned back from the wilder parts of the *fazenda* toward the big low rambling house, and as we reached the top of the last hill before descending the steep slope down to the winding river, the whole of the valley lay spread out below us and suddenly it struck me that the whole landscape was very like Pleasant Valley among the woods, hills, and rich fields of far-off Ohio. This Brazilian valley had the same soft rolling hills, the same marshes and springs and streams, the same woodlands crowning the tops and the steep slopes of the hills. The differences were not of kind but of degree. This Brazilian landscape was more lush, more extravagant, larger in scale. It struck me that the two valleys, Pleasant Valley in Ohio, and the Valley of the Atabaia in Brazil, had almost exactly the same contours, the same hills and depressions. The greatest difference was that of climate. In Ohio we had the glories of four sharply defined seasons ... the cold frost and snow of winter, the slow lush awakening of spring, the hot rich summers, and all the magnificence of brilliant blue skies and flaming trees in autumn. Here in the Valley of the Atabaia, the seasons were less clearly defined. Trees frequently flowered exuberantly in the autumn and dropped their old leaves at the beginning of spring. All the year round there were crops and always growing things. Winter and summer were marked only by a slight difference in temperature and the fact that in summer one expected heavy rains and in winter dryness.

Below us the whole of the compound around the Big House lay spread out like an aerial photograph ... the rambling house itself, almost hidden among shrubbery and flowering trees, the big cattle shed, the huge bulk of the Almaxerifado, the little lake, the great avenues of eucalyptus, the clustered white red-tiled houses.

Once the decision was made, the establishment of the farm called Malabar-do-Brasil went ahead rapidly. A syndicate was formed in which the stockholders were Dr. Santos, owner of the place, Carlito Arañha, Francisco Matarazzo, Herman de Barros, the Supermarket Corporation of São Paulo, and two or three others. The whole enterprise was to be under the direct personal direction of Carlito, whose own *fazenda*, Rio de Prata, was only forty-five minutes away on the other side of the big São Paulo–Campinas superhighway, and myself, who agreed to come down from the North to spend a part of every winter there.

All of those taking part in the enterprise were men of good will and devoted to the interests and future of Brazil. None of them belonged to the "quick-buck" speculating element which has constantly been so disastrous in their effect in all Latin-American countries and especially in Brazil ... an element which speculates constantly on the future and the vast, dormant, even unexplored natural riches of Brazil, without doing anything constructive whatever about the present.[5]

Carlito had his own big and flourishing glass manufacturing interests. Francisco Matarazzo represented the biggest industrial family in Brazil. Herman de Barros was head of a great company which has invented a kind of revolutionary pioneering in Brazil. The company goes into the wilderness, clears the land, and builds whole small towns, complete with town hall, dwellings, farmhouses, water and sewage systems, etc. When the jungle has been subdued, the company sells the houses and lands to settlers on easy terms over a long period for repayment, and the pioneers

---

[5]One of the curses of the Brazilian economy and one of the great obstacles to the development of the whole nation is a truly terrible thing known as the "*negocio.*" This is a difficult word to translate. It might be vaguely translated as a "proposition", but the "*negocio*" is usually a gamble and a speculation founded largely upon talk and the process of intoxicating the buyer. It takes many forms and manifestations, but essentially the "*negocio*" is always aimed at getting rich quick, with as little work or sound construction and development as possible. There is a whole element of the population which has no regular job but just lives on "*negocios,*" an unstable and an undependable process both for the individual and for the country. The whole of the Latin-American world is plagued by the "*negocio.*"

are spared the usual hazards and hardships of a real frontier. Of mixed German and Brazilian ancestry, de Barros represents the best of the "New Brazil," the element which in the end can make of Brazil a great, rich, and powerful neighbor.

The men behind the Supermarket Corporation represent another modernizing force which, like the revolution that Sears Roebuck has brought into Brazilian business and commercial life, is having a modernizing effect upon merchandising and commerce in general. Doctor Santos, one of the best physicians in Brazil, is a man of great intelligence and foresight, who is interested in the importance of nutrition in building a strong and intelligent people.

The team was a good one, and with energy went to work at once on the financing and transformation of the old, half-abandoned *fazenda* into a pilot farm which could bring benefits not only to those engaged in the enterprise but to the whole of Brazilian agriculture, welfare, and the national economy. The undertaking was fortunate in having the agricultural experience and the shrewd and tireless energy of Carlito Aranha.

In the background there was another good friend of the enterprise, a man of great distinction, a true patriot, a fighter for man's dignity, and the publisher and editor of the great newspaper *Estado São Paulo*, which is frequently called the *New York Times* of Brazil. This was Dr. Julio Mesquite, in reality a neighbor of Malabar-do-Brasil, who lived part of the time at the big house of the *fazenda* just over the mountains in the next valley, the *fazenda* from which Carlito, David, and myself set out on the day I first saw the Brazilian Malabar and the Valley of the Atabaia. Julio Mesquite has recognized all his life the great part which agriculture and the livestock industry play, and must always play, in the economy and development of Brazil's immense wealth. His newspaper has a special agricultural editor and an agricultural department, and Dr. Mesquite himself lives part of the time at his own big *fazenda*, very close to the problems and difficulties of Brazilian country life in all its aspects. He was one of the first to become interested in Malabar-do-Brasil and has continued to be one of the most enthusiastic and faithful supporters of the whole venture.

Finally in the planning there arose the question of the overall on-the-spot management of a big enterprise which would involve scores, perhaps finally hundreds, of workers, quantities of machinery, buying and selling, and all the complex and monumental problems of modern agriculture on any scale. Virtually against my wishes, the directors moved in my daughter and son-in-law to undertake the management of the operation. It was a great opportunity for two young people with a very small son, and one which I myself perhaps would have seized greedily at their age, but experience and years of living had made me from an adventurer into a conservative, and it seemed to me that they were undertaking a tall order for a young couple just out of the Cornell School of Agriculture.

It meant going to a new country where there was spoken a language which they did not speak, a country with a Latin culture and civilization to which they would of necessity, despite any contributions they might bring, be forced to adapt themselves. Rather sourly I viewed the whole project with misgiving, aware that save for the fatherly and genuinely friendly interest and experience of Carlito and the other directors, they were biting off perhaps a great deal more than they could chew.

Fortunately the adventure turned out well, and after a few blunders, the young people began to hit their stride. Very quickly they picked up the language ... not the classical Portuguese but the practical "*caboco*" of the village and country people; and they learned a great deal of Italian, for most of the workers came from the Valley of the Po. For my daughter it meant taking over the renovation and management of the big, rambling one-storied house which seemed to have no end, as well as a certain matriarchal position in relation to the workers and their families. All this at the tender age of twenty, with a baby less than a year old. In addition to all of this, she began before long to write a weekly newspaper column on agriculture and the problems, trials, and successes of Malabar-do-Brasil for the *Estado São Paulo*, and maintained the writing of a newspaper column syndicated in the States which had already begun a couple of years earlier from Ohio Malabar Farm. And she wrote frequently for Brazilian agricultural and women's magazines. She now has another child and her first already speaks Italian, Portuguese, and English, with a smattering of French and German as well. His favorite music comes during the Japanese hours on the Brazilian radio, set up for the great number of Japanese who have colonized large areas of Brazil. His playmates come from a half-dozen nationalities. He is already indeed what his grandfather would like him to be—citizen not of one country but of the world. Nothing is more comical than the sight of this four-year-old falling flat on his face, picking himself up and saying philosophically, "*Pusha la vida!*" which in English means merely, "What a life!" … a saying picked up long ago from the workers on the *fazenda* and in the Big House.

The history of the big farm resembled in many ways the history of many a big American farm. In a couple of centuries it had been worked out and abandoned. "The Englishman" had tried to live there as one lived in a big country house in England, in Edwardian days, with horses and dogs, a rose garden, and a small army of servants. After the good and very busy doctor came into possession of the place, with the intention of restoring it, he made mistakes through lack of experience with agriculture and livestock, many of the mistakes made in the U.S. by many a well-meaning city farmer. Worst of all he had placed the cart before the horse, as many well-meaning Americans of means frequently do. He had spent great sums of money upon buildings, fences, and installations without first doing much about the soil and the productivity of the place. And so when the old *fazenda* became Malabar-do-Brasil, the management inherited some fine buildings, some valuable heavy machinery (the most important being four caterpillar-bulldozer tractors), and a great expanse of fields and marshes that had been neglected or ignored for generations.

In many ways the fields were in the same condition that we found the fields years ago at Malabar in Ohio. The pastures were poor and thin, the hillsides gullied and eroded, the huge orchard of unprofitable tung trees the victims of voracious ants and in a dying condition. We had taken over a terrific job, but all of us were willing and even enthusiastic, principally, I think, because we understood the great fundamental and potential riches of this half-abandoned area, and knew that its restoration to productivity and profit would be the key to the restoration and real productivity of similar millions of acres in the older parts of semitropical Brazil.

I did not see the farm for nearly a year after it was taken over and the soil reclamation program set under way. I returned to it in February and found a number of prodigious changes, which would have done credit to the most energetic of modern U.S. undertakings. In some respects the place was almost unrecognizable. Most important of all, there was strong evidence in the crops, in the texture of the soil, in the obvious productive potentialities, that the farm and indeed the whole area was by no means as "worn-out" as it was supposed to be. There was nothing the matter with the soil but the way it had been farmed for a couple of centuries. In this respect it was exactly like Malabar in Ohio.

In most semitropical and tropical countries agriculture is likely to be backward because many things grow easily and without trouble. It might be said that farming in many a tropical area is simply a matter of "letting things grow." In this there is a parallel to most of our own farming not so long ago, a belief that the farmer's land "owed him a living." It is only lately that we have emerged from that philosophy of frontier-pioneer agriculture on virgin soils into a proper productive and intelligent permanent agriculture, but there still remain in the richest parts of our Middle West, farmers who still are living off the original virgin fertility, without doing much to restore and maintain it.[6]

Some of the worst farming I have seen in the U.S. actually takes place in some of the richest areas of Iowa and Illinois, where the original virgin topsoil, very deep, is still not depleted. Generally speaking, today there is more good farming in progress in our difficult areas than in our richest soils.

At Malabar-do-Brasil we had the good fortune to inherit a number of assets. Some of the buildings were magnificent and new, and beautifully and solidly constructed during the ownership of Dr. Santos. Among these were the great cattle shed, the enormous Almaxerifado, or administration building, and a handsome poultry house that could have served as a hotel, and actually later was converted to dwellings for the Italian workers. The fences over a large area were magnificent with cut posts of split granite. And, of course, the bulldozers were invaluable. In that first year they cleared fields, pulled up thousands of dying tung trees, pulled multiple plows, and made new roads and diversion ditches. Their only disadvantage was that every Italian worker wanted to operate them. Actually they were under the direction of one of the proudest and toughest men I have ever seen, a big North Italian called Vianni who looks like a *condottiere* and has a mountainous wife who seems quite able to keep him in order.

---

[6]In Brazil there is a story which I have set down elsewhere in *A New Pattern for a Tired World* concerning agriculture in the past in Brazil. It concerns the traveler who, stopping at a run-down farmhouse, found the proprietor dozing in a hammock. All around him grew wholly-wild banana plants, avocados, wild oranges, and many other fruits. Near by cattle, milk goats, pigs, and chickens roamed about with no need of shelter or feed. Rousing the farmer, the traveler engaged him in conversation and among other things asked, "Does rice grow around here?" and the farmer answered, "No." "Does corn?" Again the answer was "No." Questions concerning beans, potatoes, and other commodities all brought the same reply, "No." Somewhat surprised, the traveler said, "That's funny. I saw them growing just down the road." And the farmer, without getting out of the hammock said, "Oh! They grow if you plant 'em." This is a story which Brazilians tell joyfully on themselves.

The problem of labor was almost the first one on the list of countless difficulties. There were on the farm a few Brazilian families, but many more were needed, and the problem solved itself when one day a fine-looking gray-haired Italian appeared and said that he could deliver as many North Italian families as we desired, all coming originally from the same small region in the Po Valley. They were, like most Italians, dissatisfied with work on the coffee plantations where they had been dispersed. Some of them were ready to leave this land of opportunity and return to Italy if they could not find more congenial work and be together again, united and happy with their own people. In time, the old gentleman brought them all to the big farm in the Valley of the Atabaia—nearly a hundred families, all of them good workers, fine people, dependable and responsible and amiable. They have been there ever since, a whole small community, making an immensely important contribution to the whole enterprise.[7] The colony has its own school and the *fazenda* its own football team, which on Sundays plays the teams from neighboring big *fazendas*. To it have been added two Japanese families who are indefatigable workers and raise the vegetables which require what might be described as "personal" care. They are very good people, dependable and, of course, tremendous workers. Each time you pass their houses, a child runs out with a gift of rice cakes or some other Japanese delicacy, and when Hashado, who is no longer young, but straight and trim and tough, brings a big flat basket of vegetables to the Big House, they are arranged as a work of art, with much thought being given to their form and contrasting colors. And when he presents them, usually in the long Brazilian twilight when everyone is sitting on the veranda, and has time to admire

---

[7]Brazil has had a considerable problem with her immigration, especially during the Second World War. Coming to a new country, each nationality tended to colonize completely whole areas, in some cases taking up and continuing their new lives, as in the German colonies in the South, without ever learning the language of the country to which they had migrated. Brazil needs immigration and plenty of it to develop the country, and most of the incoming stock has made a great contribution to this development; but during the war, whole colonies of Japanese, Germans, and Italians created a problem for a nation which was strongly on the side of the Allies, and especially of the U.S. After the war an attempt was made to prevent this tendency toward mass colonization and to disperse the incoming immigrants. Immediately trouble arose, especially among the Italians, and many actually returned to Italy rather than accept the living and housing standards and generally isolated alien life on the big coffee plantations. Old Vianni and his friends who came to Malabar-do-Brasil represented this element. Once they were reunited at Malabar-do-Brasil, working with cattle, general farming, and above all with vegetables, their discontent vanished and the little colony has continued to grow, as one Italian family after another joined it.

the quality and beauty of his gift, the presentation becomes a complete ceremony, with bowings and hissings and a great deal of style.

And there are two other contractors and their families. Tom Leal, of mixed Portuguese, Irish, and American ancestry, with a wife from Northern Ireland, and Beppo Zayas, who is Spanish and married to a French wife who is the daughter of Bernanos, the poet and playwright. The fertility of Malabar-do-Brasil is not limited to the soil; there are children everywhere of every age, and each year there are many marriages and still more children. The living conditions of the workers are better than most, as are their wages, and each year the *fazenda* plants more and more of the vines and trees and shrubs which grow so easily in Brazil and provide good fruit for everyone merely for the picking ... bananas and plantains, guavas, citrus fruit of all kinds, mangoes ... indeed almost anything you can name. There are of course plenty of the troubles that go with any big agricultural operation, but on the whole it is a happy world in a happy climate where everyone can live well with less effort and less cost than in countries which are colder and damper and where the sun shines less frequently.

Over all the human and family side of the picture stands my daughter, a young American woman of twenty-three who attends all weddings, funerals, and christenings, settles quarrels, and sees that the sick ones go to a doctor and that the babies get what they need. It is very odd to hear a girl from Richland County, Ohio, being addressed by all as "Donna Elena."

For myself, Malabar-do-Brasil is a whole new life coming at middle age, and its effect is a little like that of being reborn and made young again. As at Malabar in Ohio, people ... pleasant, intelligent people ... come in daily from all over Brazil and the world. Sometimes they come for lunch and sometimes for the night or for several days. A dozen languages are heard every day, and as I write in the big white room overlooking the river which belongs to me when I am at Malabar-do-Brasil, the children outside the door are playing and shouting in Portuguese, in English, in Italian, and occasionally one hears words of French and Spanish. It is a happy world, in which differences of race, creed, nationality, or color are of little importance. Free and remote and independent among the hills that border the beautiful valley, the *fazenda* Malabar-do-Brasil is at times a small pattern of what the world may be once it begins to understand Christianity, or even true civilization.

I am very grateful to the Good Lord and to the Brazilian friends who made possible Malabar-do-Brasil, where there is never a dull moment and where everyone is working toward a common goal which may one day make a solid contribution to the life of a great, beautiful, and wonderful country.

# A BRAZILIAN FARM PROGRAM

IT MUST BE OBSERVED IN THE BEGINNING THAT THE PROGRESS MADE IN the establishment and development of Malabar-do-Brasil during the first year or less of its existence seems to me to be remarkable. In making this judgment I speak from very great experience in many countries in the rehabilitation and development of "worn-out" and semi-abandoned land and the great difficulties which arise in the beginning in the cleaning up of such farms, of establishing new and modern methods of agriculture, horticulture, and soil conservation. I visited the property now known as Malabar-do-Brasil about a year ago and considered the situation difficult, and one requiring a considerable amount of capital, a great amount of knowledge, organizational ability, and plain hard work. I can say honestly that in less than a year more has been accomplished than I expected to see accomplished in three or four years.

Not only have I been greatly impressed by the progress but also by the unexpected response of the soil itself to proper treatment. I have found that the soil is far more fertile than I had expected and that on the whole the crops grown or now growing on the farm demonstrate clearly the very great possibilities of Malabar-do-Brasil in the future, provided the project is regarded as a *permanent* one—managed in such a fashion as to restore, maintain, and *increase* the natural fertility of the soil and its potential production.

NOTE: This report was made originally after several weeks of intense study and discussion regarding the establishment and development of Malabar-do-Brasil, including consultation with the author's good friend Dr. Hugh H. Bennett, former chief of the Soil Conservation Service of the U.S., then in Brazil in a consulting capacity (1953-54).

As in the United States and all new countries, the original agriculture practiced by the earlier settlers and farmers and cattlemen is usually one of exploitation … to take everything off the land as rapidly as possible, then abandon it and take up new land. This has been the practically universal pattern in the past in the fundamentally large area of eastern and central São Paulo State. Malabar-do-Brasil obviously reached this semi-abandoned state many years ago, and since then the land has on the whole benefited by not being farmed at all. The soil has had a rest and opportunity to build itself back, simply by being let alone, which was better than having been farmed by the methods practiced in the past—a condition which unfortunately still exists in many areas of the U.S. This undoubtedly accounts for the unexpected fertility shown by almost all the crops. To consider this point in further detail:

(1) Much of the corn grown compares favorably with corn grown on completely new virgin land in the West and with corn grown on the few *fazendas*, such as Rio de Prata, where a sound and efficient permanent agriculture is in practice.

(2) The soybeans are unexpectedly good, especially in view of the fact that much of the land is in need of lime in order to get the best results from soybeans. The soybean crops compare favorably with much of the soybean production in the rich Middle West of the U.S., and is actually better than the average there. The beans in general would benefit by additional lime and phosphorus. The beans in actual seed production show a deficiency of this latter element and would unquestionably give a higher yield of seed if the element of phosphorus was emphasized. The fact that the beans have done so well despite these deficiencies indicates the general fertility of the soil, good soil structure, and the unexpected, rather high availability of the elements existing already in the Malabar fields. This availability arises from the presence in the soils of sufficient or nearly sufficient quantities of organic material, which has come about largely owing to the fact that the fields have been "resting" for a considerable period, and were allowed through root systems and the growth and decay of vegetation, and to some extent animal manure, distributed under the old half-wild pasturing program, to build up the supply of organic materials.

(3) The rice crops grown in the lower areas are on the whole well above the average in quality of growth and size of head to much of the

rice grown in Brazil. The accumulation of weed seeds through years of abandonment accounts for the weediness of the rice. As the land is kept under cultivation in succeeding years, the problem of weeds will be greatly diminished and the extra labor employed this year in cleaning the rice fields will become unnecessary in the future. The extra labor employed this year created a considerable expense, but in the long run was a good expenditure in that it prevented the ripening and distribution of billions of weed seeds which could re-create or increase the problems of weeds in succeeding years. I would recommend the use of the rotary hoe in clearing out weed seedlings from new rice, wheat, and other small grain plantings.

(4) The peanuts, which started off in excellent fashion, have proven somewhat disappointing because they were attacked underground by an undetermined fungus or blight. While I am unable to identify this disease, it is not impossible that the second and third flowerings may show an immunity and that the addition of some lime to the essentially acid soils might easily correct this condition completely. The important point is that this plantation of peanuts demonstrates that in texture and qualities, certain areas of Malabar-do-Brasil look extremely good for the production of peanuts, once the balance of elements is corrected by lime and fertilizer and the organic content of the soil increased. It should be pointed out that the peanuts were grown this year in one of the areas where the content of organic material was lowest, although the soil in other respects (structure and general looseness) was perfectly adapted to peanuts except for the slight acidity.

(5) I did not see the tomato crop, but Mrs. Geld informs me that the yield and quality were excellent, especially of the so-called "American" varieties. Judging from my own observations and from the reports, Malabar-do-Brasil should be able to achieve excellent yields of quality tomatoes. Mrs. Geld is familiar with the yields and *quality* of tomato production at Ohio Malabar. This is important in relation to the whole Brazilian market, in which the quality of tomatoes is generally low and in which the juicier, higher-flavored, so-called "American" varieties are virtually unknown. The tomato dominating almost entirely the Brazilian market is one popularly known as "Santa Cruz," a rather inferior variety of the Italian tomato-paste tomato known as "Pomadora." Greatly improved varieties of the same family are available in the U.S.

(6) I think it should also be noted that an excellent job of general cleaning up and restoration of buildings has taken place, so that in general appearance Malabar-do-Brasil already compares favorably in appearance with the best practical and modern *fazendas* anywhere in Brazil.

Owing to the fact that Malabar-do-Brasil lies in a tropical climate, but at an altitude of nearly three thousand feet with a wet and a dry season and good conditions of rainfall, the general atmosphere is salubrious and healthy on the whole for plants, animals, and people, and makes possible the raising of crops continuously throughout the year. The principal problem is that of finding the temperatures and the rainfall suitable to given crops at given seasons ... and not to overwork the soils and drain their fertility. Although habit, custom, and experience have somewhat determined this factor, there is still considerable research and experimentation to be accomplished in the whole field, especially with the increasing use of irrigation, which has made possible the cultivation of temperate climate crops requiring moisture during the cold and dry winter season. In other words, celery, which requires a cool climate but also great amounts of moisture, can be grown very well with irrigation during the winter season, when both elements are provided, where in the past it was virtually an impossible crop in nearly all tropical areas. The introduction of irrigation has produced an entirely new factor, which in many respects represents a small agricultural revolution. In other words, irrigation has provided certain areas in Brazil with a whole extra season, unknown until its introduction.

It should be pointed out that under the extraordinary conditions of the high-altitude area of tropical south central Brazil, it is possible within a single orchard to grow citrus fruit, avocados, apples, peaches, plums, quinces, pears, pineapples, and every sort of both tropical and temperate zone fruits. The same is true with the wide range of vegetables and flowers. At the time of writing both orchids and dahlias are in bloom in the Malabar-do-Brasil Big House garden. Roses and foxgloves and irises flourish side by side with every sort of tropical plant and shrub. These favorable climate conditions create for the purpose of agriculture and horticulture remarkable conditions which are extremely rare in the world and which can be utilized to great advantage in the program of Malabar-do-Brasil.

SUMMARY: The important fact demonstrated by the crops grown this year is that the soils of Malabar-do-Brasil show a very high degree of fertility, which can be greatly increased by proper specialized applications of fertilizer, the establishment of proper mineral balances, and the maintenance and increase of organic materials. In view of the fact that the *fazenda* is well located with regard to population, transportation, distribution, and markets, it has a much greater value and prospect of profits than many of the new virgin lands opened up farther west, which are remote from markets, have few and wasteful means of transportation, resulting in serious losses, and which lack proper communication with regards to markets and other economic factors. In other words, it is my opinion that the cost of restoring Malabar-do-Brasil to a level of high production and profit should be much less than that of opening new and virgin land under the handicaps of poor transportation, communication, and marketing facilities.

This is a point which should be of the greatest economic importance not only to the individual *fazendas* now in operation in the whole São Paulo area but to the general economy of Brazil. It should never be forgotten that an economy based upon the overdevelopment of industry, with a poor and unproductive agriculture, is an unsound economy which in the end will lack stability and permanence. Nor should it be overlooked that transportation and roads rank very high in the list of needs for the proper development of all Brazil.

## PLANS FOR THE COMING YEAR

The plans for working on lease contracts to individuals of much of Malabar-do-Brasil for the raising of vegetables or other specialized crops seems to me to be a good one at this stage in the development of the *fazenda*. It serves two important purposes: (1) The cleaning up and cultivation of several abandoned or semi-abandoned areas. (2) It is a help with problems of labor and management at a moment when much time must still be devoted by the overall *fazenda* organization itself to cleaning up and putting in cultivation large areas such as those now occupied by tung trees. (This large area now in unprofitable tung production contains some of the most productive soil on the *fazenda* and is greatly benefited by the fact that it has not been used for agriculture for many years. Once put into production, it should produce very high yields, especially in the areas where irrigation is possible and practical.)

The key to the success of the contract operations lies in sufficient machinery and irrigation equipment. If these are inadequate, if the ground cannot be prepared at the proper time for crops, and there is not sufficient water at the proper times, the arrangement will not be successful, the contracting growers will be dissatisfied, and the whole system will break down.

In this connection I have talked to all of the men thus far contracted for this plan. They seem to me to be on the whole men of good quality, general education, and possessed of sufficient practical experience. On the whole I would judge them superior to what could be expected in the first attempt in this direction. It seems to me that the whole success of this contract operation is based first upon the right machinery at the right time and sufficient water when it is needed. On most of the land contracted for, surface irrigation with gravity flow is practical and indeed should be easy and simple. Later on in many areas an overhead irrigation system may prove more practical and efficient and produce better results. One of the most satisfactory elements in the contract-operator arrangement is that the contractor operates on a percentage basis and is therefore possessed (in theory at least) of a proprietary interest and encouraged to increase production and efficiency.

One development which I believe to be of the *very first* importance to the *fazenda* is the establishment as rapidly as possible of a small area where tests can be made of seeds and new varieties of crops, and where experiments of all sorts in new and different ways of growing vegetables and other crops can be established. In my opinion, the best location for such a plot would be that adjoining the schoolhouse. This is good land which is certain to be high in organic material. It can be irrigated easily and is located directly on one of the main roads of the *fazenda*. This is important not only for convenience but also because it can be visited easily by other farmers and *fazenda* owners who wish to study what is being done. Similar experimental plots at Malabar in Ohio have been of the greatest possible value not only in working out problems of soil and production for the farm as a whole, but it adds greatly to the prestige of the *fazenda* and is frequently of great value to other farmers.

I have suggested to Mr. Geld that he select from among the younger members of the working force one or two boys who can be put wholly in charge of this operation, and that they be trained especially under the direction and supervision of Mr. and Mrs. Geld themselves. I have found that this is the only system by which one can be sure of having men who are not handicapped by previous conventional training in many methods which are out of date, inefficient, or downright bad. In this experimental area it would be possible to test out new crops and new varieties of vegetables and crops. The *fazenda* would thus have a basis upon which to judge the introduction of new crops and new varieties at the proper seasons without risking large expenses and losses in putting out whole acres of crops which might fail because not enough is known about them or the proper methods and seasons for their cultivation.

*I cannot emphasize too much the importance of establishing such an area, both to the economy of the fazenda and to its prestige and its contributions in general to the agriculture and horticulture of south central Brazil.*

In the area surrounding the main house and buildings of the *fazenda*, I would suggest certain changes which could be made cheaply and inexpensively, to the great advantage of the *fazenda* in many ways, especially in general appearance and convenience. I would suggest the creation of at least three more paddocks in grass and legumes similar to the one already existing next to the cattle sheds. One of these areas would be the small piece of land lying between the cattle sheds and the new trench silo. A second would be the area lying directly across the road beyond the eucalyptus trees where tomatoes were grown during the past season. The third area is the orchard behind the house.

All of these are ideally located for pigs and young livestock of all kinds. They have shade and water, either running through them or easily accessible. They are areas which would produce good grass and legumes. Three of these areas—the existing one, the one between the cattle sheds and the trench silo, and the one where the tomatoes were grown last year—could be rotated so that the young stock or pigs would always have plenty of pasture. The orchard area could be converted to raising chickens or pigs or sheep.

In my opinion, the orchard should not be cultivated but should be kept in grass, preferably with pigs or chickens running underneath the trees. I succeeded in starting this method among some Jamaica citrus fruit growers with excellent results. The yields and quality of the fruit itself increased and insect attack and even disease declined. In the cases of both chickens and pigs, they destroy and devour many of the insect pests which come out of the ground. In the case of the pigs, they also maintain by rooting, a mild cultivation which is good for the trees *without* damaging them or their roots as does mechanical cultivation under some conditions.

Surrounding the old administration buildings there is an area which offers a great opportunity for demonstration in the rebuilding of eroded and so-called "worn-out" soils. The productive topsoil of this area was destroyed or removed by the use of a bulldozer to level the ground. Here exists a subsoil typical of much of the soil on Malabar-do-Brasil and indeed of very large areas in eastern and central São Paulo State. It is the same subsoil out of which the *original* rich soils of Malabar were built up by nature over a period of millions of years. It should be remembered that all good fertile topsoil is nothing but the *original* subsoil *plus* organic material. Nature worked very slowly because she could lay down only one leaf or one blade of grass at a time, but today with an abundance of grasses and legumes used as green manure, with chemical fertilizer, and with modern deep-tillage machinery, it is possible to build topsoils at least ten thousand times as rapidly as nature.

I should like very much to see this area, which at the present time resembles a desert, built back by modern soil-creating methods. I should like to see at least two crops a year of legumes and grasses turned back deeply into the subsoil of this area. If manure from the cattle sheds could be added, the speed of the restoration would be much greater. I would suggest that the area be limed and that reasonable quantities of fertilizer be used in order to get as rapid growth, and as big quantity of green manures as possible. The slaughterhouse fertilizer from Itatiba would be excellent for this purpose. I would also like to see left untouched as a check plot a strip, perhaps five meters to ten meters wide, across the whole of the area. It is my opinion that under intensive methods this now barren area would become within three or four years one of the most fertile areas on the entire *fazenda*. This existing area offers a great opportunity for such a demonstration, of great value to the agriculture of all south central Brazil.

Eventually, once this soil is restored, the test plots and experimental garden could be transferred from the suggested site adjoining the schoolhouse to the area between the road and the old administration buildings, where it would add greatly to the general appearance of the *fazenda* and be of great demonstration value to visitors.

It seems to me a mistake at Malabar-do-Brasil to attempt putting into cultivation of any kind some of the steeper slopes ... those that are really too steep to support continuing cultivation of any kind. The one area which was terraced during the past year belongs in this category. It is really too steep for cultivation and difficult to terrace efficiently. In my opinion, these steep areas should be left for pasture where water is available for the animals, and if water is not available they should be put into grapes or fruit production, leaving the soil uncultivated and using large quantities of mulch which could be brought in from the outside or maintained merely by the constant mowing of the grasses growing between the rows of grapes. I think much thought should be given at Malabar-do-Brasil to the proper use of land for the proper crops. All of this depends upon the degree of slope, the texture and quality of the soils, and the availability of irrigation water. The entire area too steep for proper cultivation at Malabar-do-Brasil probably constitutes less than 5 percent of the total acreage.

It seems to me that the contractor called Costa can probably be of great value and any expense he represents can be justified during the first three or four years by the fact that with his help larger and larger areas can be put into production, thus increasing more rapidly the total gross return on the investment and improving greatly the actual capital value of the *fazenda*.

I think a plantation of grapes should be considered as a steady and considerable source of income. The whole general area surrounding Malabar-do-Brasil is one of the best areas for growing grapes in the whole of Brazil and even the *total* area in which grapes can be successfully grown in Brazil is comparatively small. With a steadily growing population and a rapidly increasing demand for grapes and grape products, this insures a solid investment and a good profit in the future.

The planting of potatoes should be interesting and probably profitable. The soil in the area chosen is excellent for potato culture and has a very great amount of organic material. It is possible that some nitrogen will be needed to help decay the rough organic material which has been freshly turned in, and special emphasis should be put on potash. Both potatoes and tomatoes can utilize great quantities of potash to every advantage. Mr. Geld has suggested that when the potato crop is harvested, a careful selection of seed potatoes from those hills which produce large and abundant potatoes should be made. This is an excellent idea and could easily lead to the very interesting and profitable production of seed potatoes for Brazilian use and sale at Malabar-do-Brasil. Importation of seed potatoes is expensive, and it is impossible to be certain that these imported potatoes are of the best quality and production potentiality. There is an inclination in European countries to dump inferior seed into the Latin-American areas in general, and it is probable that a process of careful selection could produce a seed stock especially adapted to Brazilian climate, rainfall, and general conditions, which would be much more productive than the seed imported from Europe. (In this connection, the more Brazil can produce for her own needs without importation, the stronger becomes the general Brazilian economy and the more stable her currency and her foreign exchange balances.)

Although I am not too hopeful about the production of wheat at Malabar-do-Brasil, I think that an attempt in this direction should be made. Considering the special soils and climate at Malabar, it is quite possible that wheat could be raised successfully. I would, however, recommend that a comparatively small area be put to wheat until the results are known. I would also recommend that small plantations be made on at least three or four different areas at various altitudes and on various soils. Attention should be given also to studying the proper times for planting, with regard to season temperatures and moisture. Wheat needs to establish a good root system and for this moisture is necessary, but once it is well established it is a crop that does not need much moisture. It is possible that in an extremely dry year, irrigation would be a benefit at the time the heads are filling out. However, it is quite possible that wheat could be grown successfully in the right seasons at Malabar without any irrigation whatsoever. (Wheat production in Brazil is of great importance to the national economy, as today the greater part of Brazil's supply of wheat is imported at a considerable loss in exchange balances.)

Certain of the lowland areas should be especially adapted to such crops as celery, onions, garlic, lettuce, and other crops of the same category, and I believe that some experiments should be made in this direction to determine the possibilities. The quality of almost all celery grown in Brazil is notably poor. As compared to celery produced in cooler climates, I believe this inferiority could be corrected at Malabar-do-Brasil by improved and specialized methods calculated to produce a cool, moist "soil climate." I would certainly stick to a definite production program for the *fazenda* and avoid the planting and production of certain crops simply on speculation because at the moment the price is high. An example of this is the plantation of tung trees at Malabar-do-Brasil which were set out on a high-price speculation basis during the war. Eventually the price fell, and now the world finds it does not really need tung oil, and the whole venture proved to be a loss. The same speculative policy has occurred from time to time in Brazilian agriculture in coffee followed by overproduction, falling prices, and in some cases ruin.

It is my opinion that Malabar-do-Brasil should establish a program of raising efficiently and at high production per acre standard crops which are steadily in demand at all times and in all years. Such a program means that when prices are low, Malabar-do-Brasil can still operate at a profit, however low, and when prices are high it could establish the highest profits per acre of any *fazenda* in Brazil. This cannot be done in a program which exhausts the soil for quick profits and which is based on changing from this crop to that one overnight in an attempt to cash in on high prices. Speculation of this kind is harmful and destructive to the economy and especially the agriculture of any nation, and Brazil in the past and present has suffered especially from this evil. (So indeed has the agriculture of the U.S. The speculative farmer is never a sound or able farmer and has an adverse effect on the general economy.)

In all tropical and semitropical agriculture and horticulture the greatest problem is the maintaining of organic material and the prevention of the leaching out of fertility to lower depths of the soils where it becomes lost or unavailable to many crops and plants. For this latter trouble the heavy tropical rains are frequently responsible, just as they are responsible for much erosion. It should not be forgotten that at Malabar-do-Brasil the oxidation and destruction of organic material takes place at least three times more rapidly than at Malabar Farm in Ohio.

At Malabar-do-Brasil the problem of organic material is one of the most important, not only because it is difficult to maintain in the semitropical climate, but because the growing of vegetables and of such crops as peanuts and even corn tends to destroy the oxidized organic material very rapidly. (All row crop agriculture in which the soil is constantly cultivated and stirred up is very destructive of humus and organic matter.) The fact that the climate permits the growing of crops the year round can lead into an agriculture of quick profits without the restoration of sufficient organic materials to the soil, and thus could result eventually in the failure of the whole enterprise. For this reason I would recommend a program of actually *raising* organic material to put back on the soils each year. This can be done by setting aside perhaps as little as 2 percent of the good land simply to grow ceresia, guandu, colonial grass, elephant grass, or other fast-growing, bulky grasses and legumes which can be harvested three or four times a year and used for mulches and composts to be put on the vegetable ground. Where corn, soybeans, and rice are grown, rotation green manure crops which provide roots, green manure, and nitrogen can maintain organic material.

I would recommend highly the purchase of a good field chopper or forage harvester as an implement which would be used almost every day in the year, (1) for making silage and bedding for the cattle sheds and (2) for the harvesting and composting of the mulch and compost crops. These forage harvesters have attachments for chopping up both grasses and corn crops.

For the creation of a constant supply of compost for putting on the vegetable and fruit lands, I would recommend the use of chopped grasses and legumes mixed with the Wilson stockyard fertilizer, or Chilean nitrate in small quantities, together with reasonable amounts of barnyard manure from the cattle sheds to supply bacteria of various kinds and other elements about which we do not yet know too much. All these materials can be layered down and the compost pile turned once during the process to mix them. This will produce a kind of fertilizer and natural soil conditioner which is invaluable and will be certain to increase greatly both the quality and the quantity of the vegetables, grapes, and fruit of all kinds, and serve constantly to increase the organic content and good structure of the soils.

I would recommend increasing uses of mulch rather than cultivation in the growing of all vegetable crops where it is possible, and if practical, even on such crops as peanuts and potatoes.

We have done much work with mulches at Malabar in Ohio and it produces the following results without exception: (1) Increase both in the yield and quality of crops. (2) By protecting the ground from evaporation, it reduces irrigation costs as much as 60 percent. (3) Mulches can reduce soil temperatures to a depth of five or six inches during hot weather by as much as twenty to thirty degrees Fahrenheit, keeping the soil about the roots cool, moist, and loose. (4) If put on in sufficient quantities it stops all but the coarsest weeds from growing and eliminates the labor and expense of cultivation, while producing at the same time better crops. (5) Mulch prevents the destruction of the small fine hairlike roots so important to the health and the growth of many plants. These are inevitably injured or destroyed by the process of cultivation, especially in the high temperature on the surface of the soils in Brazil. The destruction or injury of these fine roots tends to weaken the plants and make them subject to attack by blights, disease, and even insects. (6) At the end of the year after a crop has been harvested, very large quantities of organic material in the form of mulch remain to be plowed into the soil for the succeeding crop.

This general process approaches closely the perfect rule of a wholly successful agriculture—the production of higher yields and better quality and the reduction of costs, while constantly maintaining or actually increasing the general fertility of the soil.

Mulch is especially adapted to such crops as tomatoes, peppers, cabbage, cauliflower, etc., but can be used to advantage on almost any crop.

Water for irrigation purposes is vital to the planned operations of Malabar-do-Brasil and in some areas there is probably insufficient water for consumption from the existing streams for hours at a time. I think the construction of ponds in certain areas for the storage of water for irrigation purposes would be a good and profitable investment. Ponds are easy to construct at Malabar-do-Brasil, owing to the terrain with its deep small valleys here and there. They can be constructed with the bulldozers and earthmovers already owned by the *fazenda*. From my observations, the operators of both caterpillar-bulldozers owned by the farm are good and skilled at their work. The soil is especially adapted for building dams which will hold water. The principal expense would be the construction of concrete spillways to take care of floodwaters so that they do not destroy the dams. In some cases it would be a good idea to construct the dams with gates that could be raised and lowered to take care of floods. These ponds will contain enough water where there are running springs and streams to guarantee irrigation in quantity under any conditions. Dams could also be constructed in some of the dry ravines where there is no permanent running water, in order to trap and hold for irrigation purposes the heavy runoff water coming down these ravines during the wet and flood seasons.

One project which I think might be considered seriously is the use of these same ponds for the commercial growing of fish. Under the good climatic conditions of the valley, plus the increasing fertility of the *fazenda*, fish at Malabar-do-Brasil will grow very rapidly to a marketable size. I understand that there is a good market for fish at all times in the city of São Paulo and the surrounding areas, and that there is a constant demand for carp. Carp is the easiest of fish to grow in large quantities and rapidly in such ponds. I also understand that supplies of fish of various kinds for stocking these ponds can be obtained at the government station at Campinas. Ponds should be constructed with the idea in mind of partly draining the ponds or removing the fish with nets about two or three times a year. The large-size fish may be sent to market and the remainder returned to the ponds for further growth. Like hogs, carp can be fed grain to induce more rapid growth, and when processing plants for vegetables and fruit canning are established later on, much of the waste could be consumed by the carp, and of course by pigs in the general overall operation.

In the land along the river now in rice production, it is abundantly evident from this year's good crops that where the riceland can be flooded, much higher yields and quality can be obtained, especially if the flooding is done at the moment when the rice is forming grain. I believe that drainage ditches could be constructed at low costs in these areas, which would serve to drain the land at certain times, and by closing flood gates could be used to flood the land at other times. There is no question that this would greatly increase both yields and quality of the rice. The flooding would also largely eliminate much of the weeding and cultivation at the same time. Of course, if this system is used, it would be necessary every fourth year or fourth and fifth year to utilize the same land for dry crops so that the soil could be aerated, exposed to the sun, and the organic material and residue reduced to humus. However, such a rotation, with sufficient fertilizer and good land management, could well produce heavy crops of all kinds, and especially of onions and celery, in the off years, as well as rice. Celery, onions, and similar crops could be grown throughout the winter season when the flood gates are opened and the ditches serve as drainage ditches rather than as a means of *flooding* the same areas. Mulch would be especially valuable in the production of high-quality celery in such areas. The chief difficulty in raising celery in Brazil and the poor quality as a rule in comparison with North American-grown celery, arises largely from the question of heat and lack of steadily maintained even moisture. Celery demands moist but not wet ground. It does not grow well in soils that are wet and cold one moment and dry and hot the next. Mulch provides an even cool temperature for the roots and maintains an even moisture after irrigation. The soil does not then dry out, cake, and grow hot the moment after irrigation. Under mulch the best celery soil is a loose, open soil which will absorb plenty of water without caking and plenty of fertilizer, with emphasis on nitrogen. I am convinced that very high-quality celery can be easily and cheaply raised at Malabar-do-Brasil under the above-described conditions.

Certain areas of the *fazenda*, especially those which are too steep for proper terracing or too rocky for use, should, in my opinion, be planted to eucalyptus, either for periodic sale as a crop or for future fuel supplies in case processing plants for vegetables and fruits are eventually established. Reforestation of abandoned areas in the whole region, together with con-struction of lakes and ponds would undoubtedly have the effect of increasing rainfall. As at Malabar in Ohio, much of the rainfall (and some-times the most valuable) comes to the *fazenda* in the form of thunder-showers, which are formed simply by rising currents of hot air laden with moisture from trees and bodies of water coming in contact with currents of cold air higher up. In areas in the U.S. where reforestation has taken place and lakes have been constructed, the increase of rainfall in the form of thundershowers has been notable. Such changes, it should be under-stood, would not generally affect the heavy rains coming seasonally.

I would suggest that a dairy operation would be profitable at the *fazenda* for a number of reasons: (1) as a milk supply for the workers and their families, (2) because both milk and butter produced under proper conditions bring excellent prices in Brazil, (3) as a source of barnyard manure for the composting operations, (4) also because of the abundance of green pasture and available forage the year round.

Hogs offer good opportunities at Malabar-do-Brasil, again because of the general climate and available quantities of feed. They would be espe-cially valuable under heavy vegetable production and in connection with any canning or processing plant, as they could be used to consume all waste or inferior products.

Chickens seem a natural product, especially in conjunction with cit-rus or other orchard fruits, to consume insects and keep down the growth of grass, legumes, and weeds of all kinds. I would recommend one of the heavy breeds such as Rhode Island Reds, Plymouth Rocks, *et al.*, because after the egg production period is over the hens provide excellent chickens for roasting at very good prices. This question however is open to discussion as profits turn largely upon the price of eggs as against the sale value of the heavy hens. Leghorns will undoubtedly produce more eggs than the heavier chickens but the carcass value is much lower.

In connection with any dairy, hog, or chicken program *it is absolutely necessary to find the right men* to handle these operations. This is not merely a matter of education but equally of temperament. If the men in charge do not like and understand the animals and birds, no amount of education is of much value. In all of these operations both Senhor Aranha and myself have had great experience and are in contact constantly with all scientific advances and discoveries. Mr. Geld is also well informed.

I think there is reason to believe that the valley as a whole could become a valuable tobacco-growing area, especially under a system of irrigation and mulch, producing tobacco of high quality at very low costs of production. It is probably true that certain varieties of American tobacco would be very valuable in Brazil for producing the "American type" cigarette. We are making extensive research and experiments this coming year at Malabar Ohio in the field of tobacco under conditions outlined above, and we will be able to report by this time next year. The experiments are being made in conjunction with the tobacco growers of the state of Kentucky, the principal producers of Burley, a type of tobacco commanding very high prices on the American market. I should like to see an experimental program started at Malabar-do-Brasil to determine the possibilities of raising this valuable and highly profitable crop. Malabar-do-Brasil has the great advantage over North American growers of having available an adequate and responsible labor supply at much lower costs.

I should also like to see some seedling stock of wild or Seville bitter oranges established for the purposes of grafting, on new and better varieties of citrus fruits of all kinds. I am convinced that the system of raising citrus and other fruits on the same area with hogs and chickens can show excellent results in yields and quality far above those obtained thus far in the valley.

## SUMMARY AND CONCLUSION

In making this report and these suggestions, I have kept in mind a number of factors in the following order: (1) The establishment of a highly profitable operation for the owners, to be based not upon speculation or speculative crops, but upon a sound, efficient, and highly productive program of raising the crops, fruits, and vegetables adapted to the valley and which are in constant demand. Under such an operation the *fazenda* will make reasonable profits when prices are low and very large profits when prices are high. (2) The contribution that Fazenda Malabar-do-Brasil can make to Brazilian agriculture in general. This, I believe, can be very great, especially as Brazil has begun to pass out of the primary stage of frontier, virgin-land farming into the stage of *real* agriculture and *permanent* agriculture. (3) The need to make Brazil increasingly independent in the field of agriculture, in order to reduce the necessity for importing wheat or any food commodity of any kind, and to *increase* the production of meat and foods of all kinds for export. (4) To establish a model *fazenda* in which new and practical ideas and varieties of fruits and vegetables can be introduced, as well as new and practical machinery and methods of cultivation, storage, etc.

It would be foolish to go ahead at considerable cost in waste and expense to push production in order to get immediate results before the land and labor are ready. This means farming a great many acres to get small yields and poor quality, and nothing is more costly. In this direction (of good production and cleaning up the land) surprisingly great progress has been made, and the contract plan for raising vegetables seems a very sound and practical one under which eventually the whole of the *fazenda* will be opened up to high and economical production.

In dealing with the soil of a *fazenda* of any kind, one is not simply building a factory and then turning on a switch to get production. A good agriculture or horticulture is the process of dealing with nature and with all the laws of the universe continuously and wisely over a long period of time. To be successful in these fields requires much more knowledge and experience in more fields than in any other profession. Quick dollars and speculation are simply ruinous in the long run, and this policy of quick, immediate production with low quality and yields, is the most foolish and expensive in the world and has ruined countless agricultural projects in Brazil and elsewhere in the world.

Looking well into the future, expenditures will become necessary for a truck or trucks for the transportation of vegetables to the São Paulo and other markets, and the construction of a simple cold-storage plant in which vegetables may be stored and held when market prices are low and there are too many vegetables and fruits on the market. Such a cold-storage plant in the beginning could be constructed in one of the hillsides near the Big House, with side walls of concrete and an insulated roof covered by tile. The floor should be made of natural earth. This will serve to keep most vegetables cool and in good condition for several weeks. Later on it would be possible to equip the same storage space with an air conditioning plant which would regulate the temperatures to around 40 degrees Fahrenheit and at the proper degree of moisture. Such a plant could increase the net profits by a considerable amount, simply on a basis of advantageous prices alone. The value of such a plant would be very great in the field of potatoes, celery, all root vegetables, and even grapes and figs.

It is already evident that Malabar-do-Brasil has a great many visitors, and if the experience in Brazil follows that of Ohio Malabar, this number is certain to increase steadily as the *fazenda* develops and becomes of increasing interest to *fazendeiros* and the general public. Eventually allowance and provision for the reception and entertainment of these visitors will become necessary, not only as a courtesy, but as a valuable and cheap means of publicity and advertisement, reacting to the profit of the establishment in terms of money. This is a situation with which I have had great experience at Malabar in Ohio. In Brazil this item of entertainment is more important than in Ohio, as the distances are great and the transportation, roads, and motel or hotel accommodations are less highly developed, so that a visit to Malabar-do-Brasil becomes much more difficult than one to Ohio Malabar.

Once Malabar-do-Brasil is established as a productive base of fruits, vegetables, and other products, it would be possible to establish there canning and processing plants to utilize either the whole of the production or that part of it which would be more profitable to can or process than to sell in the direct market because prices are low. It is not impossible that such a canning and processing plant would be able in time to utilize the production of neighboring *fazendas*, and that plans of contracting with individuals to raise vegetables and fruits, such as has already been set up at Malabar, could be used on *fazendas* throughout the valley to the benefit of the whole area, in which many *fazendas* are today in an abandoned or semi-abandoned state so far as a productive agriculture is concerned. The development of canning and processing industries throughout Brazil can make a great contribution toward improving the general nutritional level, while at the same time lowering the general costs of food and protecting the economic interests of the producers in time of heavy production and surpluses. The benefits of good canned food would be especially of benefit to the residents of the large cities, where much of the population lives in huge apartment houses. Refrigerators are still largely luxuries in Brazil and quick-freeze storage is still generally unknown.

To conclude, I see Fazenda Malabar-do-Brasil as a project with possibilities of good and even high profits, having at the same time an influence for good upon the agriculture of Brazil and the economy of the nation in general. Brazil is a growing country of enormous potential wealth which has scarcely begun to be developed. It has a growing population and above all a growing middle class for the consumption of good fruits, vegetables, and other agricultural products. The advantage in the future is certain to be on the side of those individuals, elements, and syndicates which set up *permanent* production of food commodities which will *always* be in demand, and upon a basis of operation which is concerned with high production per acre at low costs and on a basis of modern efficiency. The restoration of vast areas of potentially good but abandoned land or land which is producing today only a fraction of its capacity, especially in the "old areas," could be one of the most important contributions to the present and the future of Brazil. In this, Malabar-do-Brasil could make a contribution and a very great one. Farmers and *fazendeiros* everywhere will *imitate* the operations of a practical and profitable agricultural operation, where they will often pass by the operations of a government experiment station. This has been proven throughout the United States, which at one time had, and in some areas still has, the agricultural problems which exist in Brazil.

In making this report and these suggestions I have been guided by agricultural and economic experience in many areas of the U.S., and in many tropical and semitropical areas elsewhere in the world. In closing the report I would like to say again that I have great belief in the potential fertility and production of the land at Malabar-do-Brasil, and that I have been astonished by the progress made within less than a year. The *fazenda* has many advantages over Ohio Malabar, particularly in the abundance and, to a large extent, the quality of the labor and especially the hand labor which is available.

As the project progresses and makes profits I believe that it would be excellent policy to make an arrangement for the sharing of profits on a reasonable scale with all of those *individuals in positions of authority or responsibility*, from the manager downward. This could be set up on a basis of a percentage of gross profits. This sharing system for individuals deserving it would pay many times the amount of money dispersed in this fashion and give the individuals benefited by it a sense of participation, ownership, and pride. One of Brazil's problems has been and still is the tradition of very large land holdings worked by hired labor ... a system which had been costly not only to the worker and a cause for discontent and unrest among the increasing immigrant population (and especially those who are hardworking and ambitious), but it has been costly to the landowner and the general economy and welfare of Brazil as a whole.

Again in closing I should like to pay a tribute to the interest, work, and genius which Manöel Carlo Arañha has contributed up to now in the establishment of the project. He has made a contribution which cannot be hired or purchased or even acquired by education, and which involves much more than mere business training and ability. I would also like to compliment Mr. and Mrs. Geld for the work they have done, and above all for their willingness to cooperate with Senhor Arañha and to be guided by his advice, for the general contribution they have made to the establishment of the *fazenda*, and for their understanding of Brazil, its possibilities, and its people. Also a word for Virgilio, Armando, and Vianni ... foremen and mechanic ... and all the workers and foremen—Italian, Japanese, Brazilian, French, and American—who have done much to aid in the establishment of Malabar-do-Brasil.

Louis Bromfield

NOTE: Since the writing of the first report on Malabar-do-Brasil, the progress in clearing up the old abandoned *fazenda* and in producing crops and extending the cultivated areas has been remarkable, but no more remarkable than the response of the soils to a little care, understanding, fertilizer, and improved methods of cultivation. The fertility of the so-called "worn-out" soils is astonishing, and during the rainy summer season almost terrifying. The only disadvantage is that with no severe winter season and only an occasional light frost, the range and virulence of both disease and insects is greater than in more temperate climates. However, with modern methods, these can be controlled very efficiently, with resultant big and quick-growing yields. The general program, from potatoes through tropical fruits and livestock to fish farming, has still a long way to go before it is completed, but the progress has been more rapid and remarkable than I have ever seen in any area of the temperate zones where the same process of restoration to production and fertility was in progress.

One reason perhaps is that it is possible to grow green manure crops the year round, and that the whole range of tropical and semitropical grasses and legumes frequently produces three and sometimes four crops of material, totaling tonnages per acre which would seem fantastic in temperate zones. Root systems of many of these tropical grasses and legumes are immense, and so produce within the soil to great depths huge quantities of organic material. At Aguapei, one of the enormous operations belonging to the Lunardelli family and managed by Santo Lunardelli, we traced the growth of the roots of one huge plant of colonial grass more than twenty feet downward and spreading out to a diameter of more than ten feet. This particular grass is one of the finest grasses for cattle that exists anywhere in the world, and if left unpastured attains a height of ten feet. Thousands of Brazilian cattle are fattened on this grass, and from it sent directly to market. The cattle are really fat and have the shining coats of show-ring animals. My great friend, the late Chris Abbott, one of the biggest cattlemen in the West, found it difficult to believe that the animals he saw on the big fields at Aguapei had not been heavily grain-fed. Colonial grass was imported from Africa somehow by accident, perhaps from one of the slave ships, nearly two hundred years ago. It made its first appearance far in the North in the really tropical Brazil of Bahia. A couple of generations ago some wise cattle breeder brought it south into São Paulo, Minas Geraes, and Paraná where it apparently found conditions which suited it ideally. Since then it has been planted everywhere, and within the past few years, coffee planters have begun planting it in rotation with coffee, running a highly profitable cattle business on one-half the *fazenda* while growing coffee on the other half. The rotation is indeed a long one, with a generation or more of coffee, and then an equal period of time in colonial grass and cattle, which serves to revitalize the land for coffee growing. Colonial grass has been called "the salvation of Brazil." That is a broad statement of course, but in terms of the coffee horticulture it has a great deal of truth in it.

But to return to Malabar-do-Brasil, the *fazenda*, during the year 1955 grew crops of corn, soya beans, wheat, rye, rice, peanuts, potatoes, tomatoes, lettuce, okra, onions, garlic, celery, watermelon, sweet corn, cantaloupe, all kinds of citrus fruit, avocados, apples, peaches, grapes, plums, pears, quinces, mangoes, and an almost endless range of really tropical fruit such as the jibuticaba, the chou-chou and many others. It is interesting to note that the most popular grapes in the Brazilian market today are simply the old Niagara white grapes familiar to the children of every village and farm in Ohio, Pennsylvania, and New York State. Little over a generation ago someone brought them to Brazil, where they have flourished ever since, gaining popularity until they now represent a return of millions of dollars a year to the growers. Some of the Niagaras, grown on the red and purple soils of Brazil, lost their old pale green and gold color and developed a faint pink color. There are now fixed varieties bred off the old white Niagaras which are pink, and some of them almost a deep red. For myself, I still prefer the old pale green and gold Niagaras. Malabar-do-Brasil lies in the finest grape country in all Brazil, of which the center is Jundiai, about fifty miles from São Paulo. The list of fruits and vegetables grown last year on the Brazilian *fazenda* serves, I think, to illustrate the great fertility and versatility of agriculture and horticulture over the whole of the huge plateau region of south central Brazil.

The wheat grown at Malabar-do-Brasil during 1955 turned out very well, although the area is actually too near the Equator and the climate too warm the year round for wheat. Better yields can be obtained farther south in the states of Santa Catarina and Rio Grande do Sul, which, lying much farther from the Equator and with winters which are occasionally really cold, are much better suited to wheat culture. Rye, however, gave really bumper returns on the *fazenda*. Brazil, by growing her own wheat, could greatly improve her trade balances, as at the present time she buys most of her wheat supply from Argentina and the U.S. Formerly she bought considerable wheat from Soviet Russia, but for some years now Russia has had little wheat for export without risking famine at home. It is a curious fact that although most of Brazil is ideal for growing corn and it grows like a weed even on the steep sides of hills, it is unpopular with Brazilians generally as a source of bread, although it is used for sweet puddings and by the large population of Italian descent in the form of *polenta*. On the whole, Brazilians stick stubbornly to wheat flour bread, costing the nation millions in lost foreign exchange.

Dairy operations have been established profitably on the *fazenda* with a herd of purebred Holsteins, purchased in Argentina and conducted personally to Malabar-do-Brasil by as fine a troupe of bandit Gauchos bringing their own maté as is possible to imagine. The pig program is underway, with the pigs being moved from field to field and area to area to clean up the residues of crops and to clear out some of the marshland. This system, using low fences which can be set up quickly and carried from place to place, has proved extremely successful with regard to both profits and the health of the animals. Prices of pork, and especially of lard, are high in Brazil, and these migratory pigs almost feed themselves except for some additional corn. Also the system virtually eliminates infection by worms, which is one of the problems of hog raising in Brazil. Duroc Jersey Reds are by far the most popular breed of pigs grown commercially, as they seem especially adapted to Brazilian conditions and able to stand the heat.

It may be of interest to the reader that the Mr. and Mrs. Geld mentioned in the above report are the son-in-law and daughter of the writer.

## *LA CHASSE AUX MORILLES*

IT WOULD BE IMPOSSIBLE TO WRITE A BOOK ABOUT THE PLEASURES AND the miseries of country life without considering what for me is one of the delights of living in the countryside about Malabar Farm. It is a kind of hunting which has nothing to do with guns, but requires, nevertheless, great skill, vigilance, and eyesight—the pursuit of one of the most delicate and fragile of growing organisms; something called a morel.

The morel is the most mysterious and delicious of all the wide family of fungi which gourmets consider among the most delicate and tasty of edibles. Many Frenchmen who like eating and drinking place the morel at the very top of all delicacies, above *pâté de foie gras*, above truffles, above caviar. Perhaps it is only because the morel is so rare, its season so brief, and its habits so elusive. For myself, I would join the Frenchmen who, during the brief season each year, will pay almost any price for a dish of morels.

Even among the scientists, the history of the morel is limited and knowledge concerning its habits almost rudimentary. As with the truffle, no one has ever been able to tame the morel or to reproduce the conditions under which it will grow and thrive and multiply. Since the age of four or five I have been a morel hunter, both in the United States and in Europe, and each season I find myself more baffled as to the reasons for its appearance in a given locality and the conditions in which it flourishes.

It is a fungus which grows only in certain areas of the temperate zone. It thrives in the wooded areas and the soils of what were once forest areas in the Middle West. It was not until I was nineteen years old that I learned about the exalted place of the morel in the diet of the French gourmet and that it occurs in a corresponding zone and in the same forest or near-forest conditions in France.

I made the discovery one warm day in May near Amiens during the First World War when, lazily bound on fishing, I came across a morel, the biggest I have ever seen, before or since, growing deep in the heart of a French forest. As any morel hunter knows, when one morel is discovered there are likely to be others in the same vicinity, and a search did reveal enough for a meal. But the first one was truly the granddaddy of all morels. Its bulk would have filled a paper bag of the size most hunters take with them when they set out. My fellow soldiers, who had never seen or tasted a morel, examined the precious fungi on my return with great curiosity, but most of them had never heard of eating such a thing, nor had even seen one, and the treasure was left to myself save for a bite or two *pour gouter*, and that went to the cook. Back at quarters, I lacked the cream and butter which are part of the delight of a dish of morels, but they turned out very well, cooked in a thin broth of beef which I got from the French cook of my outfit. He knew what morels were and as a reward he shared them with me. That single granddaddy of all morels made almost a meal for one person.

I do not know that the morel exists anywhere in the world outside certain areas of the Middle West and Central and Northern Europe. So rare is it that in full season it is virtually impossible to purchase a morel, either in France or in this country. In France, the supply, garnered during miles of walking through fields, forests, and orchards, goes at once to such great restaurants as Maxim's, the Tour d'Argent, or Joseph's. Occasionally one can buy a few morels by rising early and going straight to a Paris grocery shop which deals in luxuries, but one must be early for the supply, never very large, vanishes almost at once. So rare is it, indeed, that it is virtually unknown to the vast majority of people, even among those who consider themselves gourmets. Indeed, many people who live in regions where the morel grows have never tasted one and some have never seen them.

There seem to be four varieties of morels, all appearing mysteriously at the same season, which is usually the early part of May, when the weather is warm, the earth humid, and showers fall from time to time out of a sky where there seem to be no clouds. The best of all is what might be called the King Morel, which is large, meaty, and especially succulent. It resembles a sponge or a mass of brains attached to a short hollow stem. Its color is that of rich cream and in its prime it is fresh and dewy and has a delicate odor in which is embodied all the scents of awakening spring … the decaying leaf mold, the new life of ferns and wild flowers, and of forest seedlings. In itself the morel is a kind of symbol, a harbinger telling us that the frosts are over, that the spring is here to stay, that summer is just around the corner. As maple syrup is the symbol of the dying winter and the last of the snows, the morel is an assurance from nature and God that the last of the cold spring rains has fallen and the season of heat and exuberant growth is at hand. Like the bull thistle, the nettle, and the dandelion you will find the morel growing only in rich country on good land and rarely on poor and worn-out land. It loves fertility and moisture and you will find it only on the borders of rich pastures, in the virgin woods, or under ancient apple orchards.

The King Morel is found most frequently in old orchards and occasionally along ancient hedgerows on the borders of pasture or meadowland. At times it grows inside the woods, but only near the edge or in open cut-over spots where there is plenty of sunlight.

Then there is a morel which might be called the Black King, for it resembles closely the King itself in form. The difference is that the flesh of the Black King is a deep gray that is almost black. Its flavor and odor are very similar to that of the King, but it appears to be far more rare and it grows in deep true woodland, hardly ever in the open. Its rarity may be apparent rather than real, for it is the most difficult of all the morels to discover against a background of black forest soil and dark leaf mold. There is a special excitement on finding the Black King, because it is so elusive and difficult. It is to the true morel hunter what the kudu is to the big game hunter in the Congo forest. I can never remember having found more than a dozen in a whole season. Its flavor is similar to that of the King but perhaps a little stronger.

Then there is the Little King, again with a form and habit of growth like that of the other monarchs, but smaller and in color a pale gray with blackish shadows in the reticulations and depressions of the cap. It too is rare, or perhaps merely difficult to see.

Finally there is the variety growing habitually in deep woodland which bears a small brown cap at the end of a tall, broad, hollow white stem.

Among our country people it is commonly known as the dogpecker. In flavor it is not much different from the Kings and does not deserve the humble name, which is the only one I have ever heard for it, albeit the name is not one unassociated with fertility. Among country people, the difference in varieties goes unmarked, or at least undesignated, for all four varieties are known simply as "mushruins" or "sponge mushruins."

The hunt for morels is as great a passion with real gourmets as the big game hunter's search for rare specimens, and the excitement is not unrelated. I have never shot anything but lions, tigers, panthers, and Indian bison, all among the world's most dangerous animals, and then only out of politeness and with no pleasure whatever in the killing. The particular victims were dedicated to me, and considering the effort and expense of organizing the hunt and the ritual which accompanied it, it would have been in bad taste not to have done the shooting and to have pretended some pleasure and excitement in the process. For me the pleasure never came from the actual killing of the beast but from the *chase* and from the sudden appearance of a ferocious beast in the elephant grass or bamboo close at hand, or from the sudden beauty of the animal itself. All this despite the fact that some of the beasts were man and cattle killers. I cannot conceive of killing an elephant any more than I could conceive of killing one of my own children. I can say quite honestly that hunting morels is at least equally as exciting as pursuing big game. It is a much less expensive sport and, as a gourmet, I much prefer the result of the chase. Also it can be conducted within the limits of my own property.

Certainly a part of the pleasure in the *chasse aux morilles*, as the French call it, lies in the time of year and the conditions under which one discovers the morels. At Malabar we have every sort of condition in which the morel flourishes. There are deep woods which in summer resemble more the jungles of Brazil or Sumatra than a hardwood forest in Ohio, woods where there is a tangle of wild grape vines and a dozen species of ferns, where the precious and rare maidenhair is virtually a weed. Here in the deep woods grow the Dutchman's breeches, the squirrel corn, the trillium, the wood violets ranging in color from rich deep purple to a pale blue. Here too grow the yellow violets, which like deep woods with little sunlight, and the turkeyfoot, the wild anemone, the May apple, the humble cousin of the calla known as Jack-in-the-pulpit, the wild orchid, the bloodroot, and that loveliest and earliest of all spring flowers, the hepatica, with its lovely dark bronze green leaves surrounding great clusters of delicate blossoms ranging in color from white through blue and mauve to deep purple.

Your true morel hunter judges his season largely by the flowering of these wild plants. When the hepatica is in flower, it is too early. The earth is not yet warm enough, nor the atmosphere heavy with showery moisture. The bloodroot comes next, but it too flowers before the morels appear, and until the petals begin to drop it is useless to hunt for morels. But when the violets and the Dutchman's breeches and the squirrel corn begin to flower, and the wild ginger of the early settlers begins to thrust its rich lush leaves and queer, stunted dark red blossoms through the leaf mold, it is time to look for morels. When the trillium comes into full bloom and begins to turn pink, the season is over and the best one can hope to find is a stray morel, which somehow came along too late and has begun to dry up, ready to spread its spores to the wind to provide morels for other years. Then follows the hot, rich verdant summer and the brilliant ... brilliant almost to harshness ... autumn with its azure blue skies, its barbaric forest colors, yellow for beech and gum tree, purple for oak, red, yellow, and green all at one time for sassafras, and whole pandemoniums of color and glory in the maple. After that the long cold winter when all the trees stand dark with a bleak Spanish beauty against the gray sky and the drifts of snow. All that time one must wait for the return of the morel and the superb resurrection of which it is a symbol.

At Malabar there are beyond the woods the richest and deepest greens of the pastures and meadows, the ancient orchards where still grow some of the descendants of Johnny Appleseed's first trees. And there is the Jungle, a low swampy area filled with springs and with a musical spring stream flowing the whole length of it. Here one finds acres of violets—white, yellow, blue, and purple—and the spotted leaves and the yellow blossoms of the lily commonly known as the dogtooth violet. Here too there are great areas yellow with the rich gold of the marsh marigold and rank with the deep green tropical foliage of the skunk cabbage.

Here in the Jungle, when looking for morels, one follows the tracks made by the (deer, who live well in our country with its forest and marsh cover and its rich fields of alfalfa, ladino, a dozen grasses, and plantations of grain of all kinds. They are big fat deer and there is no season on them. At morel time, when the cattle have first been turned out to pasture and cavort across the green clover and grass like calves, with tails up and legs kicking, the deer gather in the late evening and the early morning about the salt blocks.

Certainly part of the pleasure of hunting morels is the sudden coming upon a herd of deer ... a big buck who has driven off the other bucks, four or five females, and perhaps three or four of last year's fawns. If you are downwind, they do not notice you at all and you can watch them as long as you like. Then at the slightest noise, the merest crackling of a trampled broken dry twig, they will turn in all their full beauty and watch you for a moment before bounding away across the meadow and over the highest fence of wire, more with an air of bouncing or floating than of running, with their white flags visible among the far-off trees of the woods long after the tan and deep brown of their coats has merged with the dark colors of the spring forest. Such a moment is for me one of the very glories of living, so filled with beauty and glory and even of love for all living things that the heart nearly bursts.

Then they are gone and the search for morels begins again.

And sometimes one comes out of the Jungle or the forest into the rich bottom pastures where the bluegrass is already deep and green and the watercress thick in the springs and the marsh marigolds in full bloom. These are alluvial lands with the piled-up fertility swept by floods from the upper valley, in the days when farmers were careless, ignorant, and extravagant. Here in these bluegrass bottoms one finds, never the dog-pecker or the Black King, but *the* King and sometimes the Little King.

And there are the wonderful smells, and let me point out that one can be a gourmet at smells as well as at foods; indeed, no true gourmet can exist without an acute and sensitive olfactory nerve. There is first of all the musky scent of leaf mold, damp with new rain and warm with the spring sun, and there are the faint delicate essences of the new ferns just pushing their way through the leaves (these too are good to eat when they are young and succulent and tender). And there are the various scents of aromatic plants, the faint pleasant smell of the first pennyroyal, and the lovely odor of mint crushed underfoot, an odor associated in my mind always with morel hunting and with bass fishing, for in our Ohio country every bottom is full of mint along the edge of the clear swift-running streams where the smallmouths flourish. And now and then downwind one catches the smell of fresh earth where a vixen fox has dug beneath the rocks a den for her cubs, and the stronger pungent smell of the dog fox lingering nearby to bark his warning of our approach. And now and then downwind comes the smell of a skunk out foraging for grubs in rotten tree trunks and the fresh newly thawed earth, and again, late in the season drifting downwind, the thick, rich perfume of the wild crab apple, one of the loveliest of all trees, with blossoms that shade from the deep red of the buds through pink to the white of the fading blossoms. And there is a peculiar dry and pungent scent that comes from the thickets of wild grapes, where the first buds have begun to show.

Proust says that the sense of smell is the greatest evoker of memories among all the senses, and in this he is probably right. All these scents are associated in my mind with my own Ohio country and the wild, quick, intoxicating spring which rushes riotously into summer. Above all they are associated with the hunt for the morel. That day in May in the heart of the Valois, in France, when I put the big morel I had found close to my face and smelled it, I was transported at once three thousand miles to the other side of the world. For a second or two a homesick soldier was back again in the thick and fragrant woods of Ohio in early May.

In our country almost everyone is a morel hunter. With the coming of the season not only the farmer takes to the woods and fields in search of the elusive fungus; in the later afternoon almost every woods in the county has a car or two parked at its edge on the roadside. Deep inside the woods or the marshes someone is hunting the morel, hunting with excitement and the most wonderful, innocent pleasure—the pleasure of being glad and thankful to be alive.

There are old gentlemen and ladies who go softly and gently hunting morels each spring. Some of them have been hunting them for fifty or sixty years or more. It is a gentle sport at which one can take one's time, choosing the slow rather than the steep climb, the roundabout rather than the direct abrupt way, for one never quite knows when one will come upon a "nest", and as in life, the gentle way is better than the steep, hard, violent one.

Naturally among so many veterans there are countless theories of how to find morels and where they grow best and why their ways are so erratic and mysterious and why one year they will pop up in one place and in another year in quite a different place, so that although you may know a woods or pasture like the back of your hand you can never be sure on your return of finding the morel. I have known spots which have gone on producing morels year after year and others where in one year there were great nests of them and the next year nothing at all.

Each year you develop new theories. You begin to think that instinctively and by the "feel" of things you know good morel territory where, in theory, they should be thick on the ground. Then, unexpectedly, you will find an unruly morel or even a nest of them growing where no proper morel should grow, and the territory which should be spotted with them is quite barren, and you are forced to revise all your judgments.

Some things the true morel hunter does know ... that abundance follows winters when there has been much rain and snow and the ground is filled with moisture to the point of wetness. Yet if the spring is slow and the rains cold, the hepaticas, the bloodroot, the wild ginger, the trilliums, and all the others will come and go with scarcely a morel to be found. There must be moisture *and* heat almost as strong as that of midsummer.

The hunter knows also that old orchards are good for finding the King, that the deep woods are the haunt of the dogpecker, and that the Black King and the Little King may be found almost anywhere there is fertility or virgin soil. He also knows that all morels seem to like the vicinity of ash trees, whether in the forest or on the edge of a meadow, and instinctively as he walks through the still almost leafless woodland, his eyes search for the peculiar, finely reticulated bark of the ash, coated slightly by the whitish gray fungus which, like the morel, seems to have an affinity for ash trees. A true morel hunter can spot and recognize the dark trunk of the ash at a distance of a hundred feet and begin to move toward it, never, however, taking his eyes from the ground lest he miss a single fresh moist morel among the dark decaying leaves.

The true hunter knows too that the morels appear to have some liking for the vicinity of the tangles of wild grape vines and on the edges of great clumps of blackberries where the old canes are decaying slowly among the fresh green blades of thin forest grass. And he knows too that the morels will appear first and most abundantly on slopes which turn toward the south and the southeast. These are the slopes which warm up first and get the early morning sun of spring. It is possible also that in our country, morels abound on such slopes because the prevailing winds are from the north and northwest and carry the flying spores over the crest of the hills to deposit them in the eddies and quiet pools of air which exist to leeward.

To be a good hunter, one must have a *seeing* eye. Many a time I have spent a whole morning or afternoon with a novice hunter who found only one or two morels, as against my own rich harvest. He could not *see* them even when he stood in the midst of a veritable "nest", unless they were dogpeckers whose great white stems at full maturity made them clearly visible. At the prime stage, the morel thrusts its delicate moist bisque and brown head through the brown and tan leaves. Sometimes you can trap one merely by observing that in the leaf mold there is a slight hump which indicates that beneath it something is growing and pushing its way toward the light. Moving aside the leaves one finds a fresh morel, or sometimes only the bishop's crosier of a fern or the firm strong stem of a snakeroot bud. But your expert hunter can tell the difference in the hump raised by a morel and that raised by common vegetation. I cannot tell you why. Again it is the *feel*.

The real hunter goes into a kind of trance in which his eyes become adjusted with so great an intensity and delicacy that they can distinguish a fresh morel pushing through the leaves at a distance of twenty-five feet or more. The hunter's thoughts may be far away, but his eyes go on *seeing* with a supersight. It is not a matter of intense concentration; I have seen novices with faces contorted by concentration who could not see a morel at their very feet and sometimes anguished my soul by stepping upon one. I think the trance has something to do with an inner passion and excitement and with the degree to which the hunter is keyed to all things in nature. There are some people who could never learn to find a morel and there are some I have seen who made a great haul the very first time out. One of the great pleasures of the hunt is that the eye, relaxed but vigilant, works while the mind is elsewhere, busy with other things. Then suddenly the eye sees a morel and a bell rings and excitement floods both mind and body.

Malabar is a large farm with a great number of men, women, and children living on it. Many of us are morel hunters, almost by profession, but rarely do we go out in groups or parties. We go *alone* or at best with one other person, to find our own "nests" and mark them for another year with the greatest secretiveness. Now and then some scoundrel follows to spy, and on one occasion when I had made a great haul, Kenneth Cooke, with the skill of an Indian guide, retraced my steps until he found the "nest" with the bare white bases from which the morels had been picked. Even the most honest and truthful of men will turn liar when questioned on his return with a great catch of morels.

The solitude of the chase is one of its greatest delights, especially for one whose body constantly demands physical action and which cannot relax into dreaming while lying on a grassy bank. In a life which is fiercely active and constantly filled with people and human contacts, the mere relaxation and subdued excitement of hunting for morels, even though there is no bag, provides exercise and solitude and time and mood for reflection. If I have troubling problems annoying me, the surest way to solve them or to be rid of them is to set off with dogs on a hunt for morels.

The dogs themselves love these excursions and for every mile I walk they run ten, following the exciting scents which crisscross the damp warm woods and fields like a spider's web. They will run off out of sight and be gone for ten or fifteen minutes and then circle back to check on me, panting with their tongues out, full of delight and excitement, trying very clearly to tell me that they have come upon a herd of deer or run a fox to earth or sometimes, when no words are needed, that they have discovered a dignified skunk who refused to run.

The morel hunts provide the best of all solitudes—not that of a close and austere cell, nor of a musty library, but of the open air, of the birds and the trees, and the sudden lovely view of a rich and beautiful valley or a ravine with a brook in spate breaking into foam over ledges and beds of gravel. It is the time of year when the mourning doves return and the wood ducks have just begun to make their absurd nests high in the old and dying trees along the creek. One sees the first sudden, brilliant flashes in flight of the scarlet tanager, which is a shy bird and likes only the thickest woods. And now and then from overhead comes the raucous obscene cawing of that mischievous and supernatural bird, the crow, which lives always in communities with its own sentinels and spies. And in the meadows the cock pheasants, in the first brilliant plumage of mating season, start up from under your feet with a sudden shrill cry and a noisy rush of wings. The red-winged blackbirds with their bright song and smart red and black plumage are in the marshes and the first bobolinks have appeared in the meadows, with the quail and the little song sparrows, which would rather run in and out among the hummocks of grass or along the lanes than fly upward into the bright air.

Morel hunters have their own superstitions. My principal one is never take a bag or a basket in which to place the catch. The same superstition has haunted me all my life as a fisherman. A creel is not a part of my equipment and the fish I take are strung on a willow or birch twig. I feel profoundly and irreconcilably that if I go prepared for a great catch, I shall take nothing at all. My only morel hunting equipment is a bandanna handkerchief which, by gathering together the four corners, becomes a proper morel hunter's receptacle. If I make a great haul, I take the shirt off my back and gather the morels into it; but the occasions have been rare when I have had to take off my shirt and when it happens, it is always a day to remember and one of the high spots in a lifetime.

The morel chase is altogether different from the hunt for the common field mushroom of the late summer and autumn. These too abound in our pastures but they are easy prey, standing out white against the short green grass and their flavor is far less delicate and wonderful than that of the morel. Indeed at Malabar we have reached the degenerate stage at which we go hunting for field mushrooms in a jeep. One can drive speedily over the pastures and spot them at great distances, without either mystery or exercise or a sense of triumph and excitement. You cannot take a jeep or even a horse into the thick forest and the swamps and Jungle where the morel flourishes best. Besides, man has learned to tame and grow in boxes and cellars and caves the common field mushroom. No one but God can produce the morel with its delicate and special flavor which might be well described as the apotheosis of May.

The people who live in my country have calves to their legs, and arches in their feet as do all people who live in hill country; there are no pipestems, nor any piano legs, and the morel has produced many a handsomely turned limb among both boys and girls, for the walking of the morel hunter is not only up and down hill but it is over *rough* land up and down hill. There is no better exercise and no more diverting exercise. For my money nothing is more deadly in terms of boredom than exercise for the sake of exercise and I suspect strongly that those people who lift barbells and perform on gymnastic apparatus and do pushups in the morning and evening, do so out of some narcissistic impulse which betrays itself in the fact that so many of these people work before mirrors and are mirror worshipers. Exercise arising from a purpose such as a contest of skill or from an impulse to create, whether it be a garden or a canal or a stone wall or a small mountain of hay, produces the same results with far greater pleasure and satisfaction. Of course, it could be said that the professional exerciser is building his own body, cutting down weight here and building a muscle there, but he risks in the end a terrible thing … a muscle-bound body which must approach in beauty the statue of Michelangelo's David in order to justify the expenditure of time and effort. And there is a still more terrible penalty … that he is condemned throughout life into eternity to doing pushups and lifting weights to keep off fat and maintain the huge muscles he has created.

Once you have made your bag of morels—and sometimes the bag is small—you start for home across country, tired, happy, and with a voluptuous anticipation of eating them.

There are a dozen ways of preparing them. There are ways in which they can be made to stretch so that a small catch will seem a big one, and that special flavor contained in the juice and texture of the "mushruin" can be made to go a long way and flavor foodstuffs which are not so delicate nor so special. They are excellent cooked in the juices of a thick steak and served up with it, and when the catch is small, they may be made into a cream soup which has no rivals. Perhaps best of all they can be cooked in butter or in a mixture of cream and butter with no flavoring but salt and pepper, for one is unwilling to risk any process which might damage or alter the rare and delicate flavor.

Then suddenly at the end of the season when the May apples begin to cover the damp earth with a thousand small umbrellas, each having for a blossom a single waxy camellia, and the trilliums have begun to turn from white to pink, the morel will vanish as quickly and as mysteriously as it appeared.

Sometimes I have worried lest in the heavy hunting that goes on in my country, all the morels in one season might be found and there would be none left to continue through the marvelous law of growth, birth, death, decay, and rebirth. But always there seem to be enough morels left behind, concealed beneath the May apples or in obscure undiscovered spots, for the continuance of the delicacy. For generations, since the first settler came into the country, morels have been the subject of a fierce, greedy, and exciting hunt, but year after year they seem to persist, rare in the bad years and abundant when there is moisture, showers, and hot sunlight at just the right moment. The lovely, delicate little fungus, each one a small monument of beauty, structure, and flavor, seems clever and mysterious enough to carry itself along, despite the most relentless hunting, year after year and generation after generation. Anyone who likes eating and who knows the morel may well be thankful to a God and a Nature which has produced and maintained the "spring mushruin."

## A SOMEWHAT TECHNICAL CHAPTER FOR ALL WHO
## LOVE AND UNDERSTAND THE SOIL

THE MARKET GARDEN BUSINESS AT MALABAR ACTUALLY BEGAN FOURTEEN years ago without our knowing it. Certainly we never had any intention of going into the business of producing vegetables for sale; the whole project, you might say, "snuck up on us." It began first with the business of producing good vegetables for the consumption of all the families living on the farm. It was a part of a program of self-sufficiency, which paid off wonderfully during the war; and could pay off wonderfully again under the same circumstances. Throughout the war all those of us at Malabar had good meat, milk, cream, eggs, butter, and unlimited fruits and vegetables. Even the shortages of such things as sugar did not affect us much; we acquired what sugar we could on rationing and the rest of the need was made up with honey and maple syrup. In some ways we gained, for we discovered, among other things, many an excellent recipe which turned out better when made with honey and maple syrup than when made with ordinary sugar.

Following the war, we modified somewhat the whole program of self-sufficiency, a process which has been described in the earlier books on Malabar. We found simply that we were engaged in doing too many things and that some of these things became uneconomic in the light of modern communication, transportation, and specialization. It was cheaper, for example, to buy our eggs and chickens from the neighbors in the poultry business, than to attempt raising and producing them for the great number of men, women, and children who lived at Malabar.[1] The man from whom we bought eggs and chickens bought milk and cream from us. Both of us produced our respective products efficiently and economically because that was our business, and more and more, as the pattern of "general" farming begins to disintegrate, it becomes increasingly evident that a farmer should be in a given business or out of it, that he should specialize and do well and efficiently a few things rather than attempting to do too many things inefficiently and with almost overwhelming costs of varied machinery and equipment.

But in the vegetable and fruit department we stayed in business after the war, even when we could perhaps purchase some of the vegetables and fruits at less than it cost us to raise them.

There were a number of reasons for this, most of them derived perhaps from myself. I am and always have been, in the whole field of agriculture, primarily a "soils" man. Economically and philosophically I believe that good and productive soil is the very basis of man's existence, well-being, and prosperity. Certainly it is the very foundation of the average farmer's success and prosperity. One of my earliest memories is of standing in the

---

[1] We have a project scheduled for a return to poultry production but upon an entirely new basis, or perhaps the oldest basis in the world. Instead of raising chickens and producing eggs under the high-pressure methods now prevalent almost everywhere, with the chickens enclosed the year round and fed upon a high-pressure diet, we mean to house the chickens in small operation units of two to three hundred, in shelter houses constructed of aluminum with thermopane glass, which prevents the passage both of heat and of cold, with an overhang roof which shades the interior of the house in summer when the sun is overhead, and admits the sun in winter when it moves low across the horizon. These houses will be located in areas near barns and yards where the chickens will have good alfalfa-ladino range in summer and fresh cattle manure will be available the year round. The birds will be allowed their freedom at all times, winter and summer, and we shall follow the ordinary practice of the farmer's wife who keeps the money in the old sugar bowl … that of throwing so much grain on the ground every day. From preliminary experiments, we are inclined to believe that under such a system, we shall avoid virtually all of the poultry diseases and, while our egg production may be lower than that of hens raised under the high-pressure systems, the costs will be less than a third that of the majority of poultry and egg producers today. One hears constantly of middle-aged and elderly people retiring and going into the poultry business, and in the great majority of these projects, failure is the result, simply because of the high cost of complicated grain and concentrate feeds and the high incidence of diseases and cannibalism produced by artificial conditions, pressure feeding, and concentrated production. Actually, today the only money being made in the poultry business goes to the farmer's wife with a small flock which runs free or to the producer with thousands of birds who buys his feed in carload wholesale lots and operates in the most efficient manner possible. No market for farm products is more unstable and more erratic than that for poultry and eggs, and the only safeguards against it are low costs or assembly line efficiency.

garden of my grandfather's farm and experiencing a feeling of distress, almost approaching sickness, at the sight of the "poor spots" in the fields of corn wherever there was a slope or a knoll and the soil had grown thin. I wanted the whole of the field to be deep, rich, and green, and I think that marked the very beginning of a lifelong passion for gardening, farming, and soils. I could not have been more than six years old.

If there is one single feature at Malabar in which I take great pride it is that on about a thousand acres of hilly land there are no "poor spots" in any of the fields; the knolls, the slopes, even the steep hillsides, all produce evenly and well, and each year this becomes increasingly true as the fertility of the soil continues to increase.

I like the cultivation of vegetables and flowers because it represents an intense application of all the rules of good agriculture and good agronomy, and because the heavy production and drain upon the soils made by heavy production per acre provide the most acute of all tests. One can watch the plants and know intimately what they like and what they do not like, what produces health and vigor, flavor and quality, what produces slow-growing, woody, flavorless vegetables, and what produces vegetables that are quick-growing, vigorous, succulent, and resistant to the attack of disease and frequently even of insects. It is a concentrated way of becoming thoroughly acquainted with all the conditions and processes existing and taking place in soils. It is like farming under a microscope.

It is true that one can have as much excitement and satisfaction out of a fifty-foot-square plot in a backyard as out of fifty thousand acres. Indeed, if you are a good farmer or gardener, you can find more excitement and satisfaction out of such a small well-managed plot than many a giant wheat farmer can find out of thousands of acres of mediocre wheat production ... certainly vastly more satisfaction than the low-production wheat speculators who go by the name of "suitcase" farmers, the men who have cost the nation so much not only in taxes but in the destruction of land through erosion by wind and water. These "suitcase" farmers seem to me to rank among the contemptibles of our time, somewhere between the Capones and the money lenders practicing usury. They belong to a breed with which, fortunately, I have had little contact, and have less and less as time goes on. I am an easygoing and tolerant man, but one has to draw the line somewhere.

But it was not only soil and the mysteries which go on within the earth which kept me producing vegetables; it was also that I like to eat and to eat well. I like not only good recipes and a great variety of dishes with plenty of herbs and spices and rich sauces, butter and cream, but I like them made out of first-class materials ... clean, succulent vegetables and fruit high in vitamins, in minerals, in flavor and freshness. Increasingly it becomes difficult to obtain this quality of vegetables and fruit. Eating in the U.S. has become increasingly a matter of convenience and labor-saving rather than of good materials or skill in cookery. (But all that belongs in another chapter.)

There are so far as I know only two ways today of having the top quality in fruits and vegetables ... either to grow them yourself or to drive into the country and buy them fresh from the grower, not from one of those roadside stands which buys its supplies wholesale in the city, transports them, wilting and even rotting, to the country roadside stand and then sells them to you in the deceit that they are "home-grown." Many, if not most of the so-called "roadside stands" today operate on such a method of procedure, and frequently the quality of the vegetables and fruits which you buy at such a stand is lower than that to be found in the chain store city market, where at least refrigeration has played some role.

And so because I love soil and good eating, the vegetable gardens at Malabar have been maintained at full production from the very beginning. I like my peas, my sweet corn, my young string beans straight from the garden into the pot. I like my tomatoes sun-ripened and still warm from the sun. I like my potatoes grown in the best of soils, so that not only are they clean and beautiful to look upon, they are also full of flavor, as different in flavor and protein content from potatoes grown in fertilized sand as day is from night.

But there was another element in the maintenance and development of the vegetable gardens at Malabar which crept in almost at once. That was concerned with the growing prevalence of disease and insect attack which one found everywhere in the U.S., and worse, the constant, universal, and increasing use of inorganic, chemical, poisonous dusts and sprays with which nearly all the vegetables consumed are constantly sprayed, dusted, and drenched ... vegetables consumed more or less directly from the field, and vegetables that are quick-frozen and canned.

One of the first things I discovered on returning from eighteen years of living in Europe was the fact that the garden pages of newspapers, the seed catalogues, the agricultural magazines, and of course the pages of the chemical trade papers were filled with a constantly increasing array of high-powered inorganic chemical poisons. After a brief survey one was led to believe that it was impossible to raise any crop in the U.S. without saturating it at some time during its period of growth with all kinds of violent poisons.

All this came as a surprise to me for a number of reasons. One certainly was that in my grandmother's vegetable garden and even the vegetable gardens which I had set up myself as a boy, all this array of violent poisons was unnecessary. There was virtually no disease and the attacks by insects, certainly of some species, were negligible. Another was that in all the years I had farmed in a small way and gardened in Europe it had been unnecessary to use anything to protect the growing plants but ordinary hydrated lime to prevent the snails and slugs from devouring a small range of vegetables and flowers. And after I returned, I discovered another fact which was of great interest. It was that in our valley, where some of the vegetable gardens near the old farmhouses were a hundred years old or more, the farmers' wives found no necessity for the use of all sorts of dusts and sprays. Plants seemed to grow, healthy and resistant and vigorous, although in the fields of the very same farm the crops were frequently poor and sickly.[2]

All of these facts plus many other minor ones immediately led a naturally skeptical and questioning mind to do some speculating. I could not help wondering about the great change which had come over the soils and products of the nation as a whole. It was not merely that certain insects and diseases, sometimes new to this country, had been introduced from time to time. (Something like the chestnut blight had wiped out whole species which had no immunity or pathological resistance.) That was understandable on one level of science, but there were many things which were not understandable, especially among the large growers who, in theory at least, were provided with all the latest scientific knowledge

[2]After watching and studying the gardens of the old farms in our valley, the reasons why the average long-established farm garden seems highly resistant to the attack of insects and disease have become increasingly evident. They are simple enough, merely that there is abundant organic material and bacterial life in these old gardens and that they have a good balance of minerals, frequently in organic form. All of these factors have been created by the annual use on these gardens of large quantities of barnyard manure. In other words, the soils of many of these old gardens have almost the texture of compost, and a large part of the fertility of the whole farm in the form of its products, both forage and grain, has been transformed into barnyard manure and concentrated upon these gardens year after year for generations. While in the case of some farms the fertility of the fields has gone steadily downward through the use of insufficient manures, both green and barnyard, and a general depletion of organic material has occurred through the excessive production of "cash crops," the gardens at the same time have been fed lavishly and have constantly grown more productive and more resistant to the attacks of both disease and insects.

regarding the inroads of disease and insects. On the whole, these large, professional, commercial growers were frequently the greatest sufferers from disease and insect attack and were the greatest users of violent inorganic chemical poisons.

I am an honorary life member of the National Vegetable Growers Association, and I am familiar with the problems, the skills, and the attainments of the commercial vegetable and fruit growers from one end of the nation to the other. During the years in which I have been associated with the commercial growers, one factor after another seemed to emerge. Within the ranks of the growers there are operators of every kind and sort, producing vegetables in all the great variety of climates and soils that exist in the U.S. Some are excellent scientific producers; some are successful and practical small operators; a few operate upon a vast scale in which the production of vegetables and fruits ceases to be an agricultural and horticultural operation and becomes increasingly a kind of industry. Some of these have fallen into great trouble, and sooner or later are faced with very serious problems of production and quality. It is within their ranks that, largely speaking, disease and insect attack have made the greatest inroads and the most widespread use of dusts and sprays has become necessary, if any crop of even a mediocre quality is to reach maturity.

There are perhaps a number of fairly obvious reasons for this. (1) Agriculture and horticulture are *not* industries and can never be treated as industrial operations for long without risking disaster. (2) The tendency in the past among very large vegetable growers has not been to look for the best and fundamentally well-supplied and balanced mineral soils, but instead to seek soils which by their very mechanical and physical texture were easy to work. The soils which are the easiest to work are not necessarily the richest or the most productive soils. Most muck soils and soils with a high sand content provide good mechanical mediums *in which* to grow vegetables, but it is rare indeed that such soils are not unbalanced, minerally poor, and frequently actually deficient in many of the minor elements important to the growth and nutrition of plants, animals, and people.

The muck soils, while they may be easy to work, are frequently almost entirely composed of nitrogen and carbon, and require the constant use of fertilizers of wide variety in order to produce mature plants or healthy animals and people. Soils which are actually peaty in quality are not only poorly balanced, but they are subject to rapid oxidation and deterioration when repeatedly plowed and cultivated. In many areas, after years of operation, these soils presently disappear altogether; they have simply become oxidized out of existence, a process which cannot happen with truly good and well-balanced soils, since such soils are composed not mainly of carbon but of a wide variety of the *actual minerals* which feed plants, animals, and people.[3]

Among friends who have been using mucklands in other parts of the country, much of their land has now oxidized down to the shale, almost impenetrable, which centuries ago trapped the water and created the marshland that in turn over thousands of years produced the deficient, peaty muck soils. Once the muckland has been oxidized or has blown away, the remaining land is virtually valueless for any agricultural production since the underlying shale is of such poor quality as productive land that it is economically impossible to restore it or convert it to production. Occasionally in muck and peaty lands fires become started accidentally and the very soil itself burns away. Frequently these fires occur underground and are difficult or impossible to arrest, and burn for weeks and even months. In our own area in Ohio, these fires occasionally occur in mucklands, and for days and weeks the air for a hundred miles around is filled with smoke arising from these underground fires.[4]

Soils with a high sandy content are as deceiving and as disillusioning as muck and peat soils when it comes to sound production of quality vegetables. Such soils are easy to work, but fundamentally they are always deficient soils, since the fertility and even the chemical fertilizer constantly applied to them to "manufacture" or increase their fertility leach rapidly downward with every heavy fall of rain. One has only to pour water over sand to understand this destructive and damaging process. In

[3]On the author's first visit to the drained swamps of the Everglades area in Florida, where vast amounts of winter vegetables are now grown, the local county agent observed, "You are not going to see agriculture here, but only a form of hydroponics in which fertilizer and minerals of all kinds must be added every year or two in order to produce any healthy, decent vegetables." The same is true of many of the sandy soils of Florida, where real success and profits are determined not by the soils themselves but by the skill, knowledge, and experience of the individual vegetable or fruit grower. Many of these Florida soils produce high gross returns but they require heavy annual expenditures and skillful use of fertilizers to maintain their production and general fertility.

[4]One of my earliest childhood memories is of oil country in southern Indiana, where a combination of oil wells, flares, nitroglycerin explosions, the distant glow of peaty, muckland fires, and vast clouds of smoke, plus cheap and abominable hotels and food and a hillbilly population created a general impression of something out of the *Inferno*.

all too many cases, the vegetable grower has been drawn to such lands because they are *easy* to plow and cultivate, and both muck and sandy lands *seem* to be naturally adapted to the intensive growing of fine-seeded vegetables and certain fruits such as strawberries. The fact is that this shortcut *easiness* is a delusion and that many a heavy clay soil, with the fundamental mineral balance and "meatiness" of soil will produce much finer vegetables and fruits in the long run than the shortcut sandy or muck soils, and in the long run at a much lower cost.[5]

The principal factor involved is that of organic material. Muck soils are usually little more than pure organic material and nothing serves to answer the organic material crank more thoroughly than the fact that these highly deficient soils *are* pure organic material and little else. Their deficiencies can be corrected in only one of two ways ... by the application of dolomitic limestone, greenstone (decayed granite high in potash), and phosphate rock, which is the system of the organic extremists, or by the intelligent use of chemical fertilizer containing the missing elements, major or minor, or both. One could compost the vegetation grown on these deficient soils for hundreds of years but in the end one would still have only muck soil with all its deficiencies. Many of the mineral elements must be artificially incorporated into such soils in order to produce reasonable quantity production in terms of economics, and certainly any degree of quality at all in terms of flavor, vitamin, or mineral content. Unless the bacterial and moisture content of such soils is consistently very high and balanced, the first process—that of employing basic fertilizers in rock form—becomes uneconomic, because it is too slow to bring an adequate return on the capital investment. Once production and fertility are established, all fertilizers in primitive rock form gain in value as factors in the maintenance of fertility and balance, but frequently in the beginning this method is too slow. In other words, the difficulty with using the "natural" rock materials is that the process of breaking down such materials and the long wait for their availability in amounts sufficient to show reasonable progress in production

[5]The fact that certain gravel, loam, and clay soils produce much finer vegetables and even higher yields than muck and sandy soils is just beginning to be recognized, and in our Ohio country some of the wiser muck-soil growers are beginning to move out of their areas into the alluvial glacial valleys of northeastern Ohio where the fundamental factor needed to make these soils highly productive is simply the addition of sufficient quantities of organic material to improve soil structure and set in motion the bacteriological processes which lie at the very base of any good, sound, quality agriculture. It should also be pointed out that the actual effect of purchased chemical fertilizers on such soils is not only immediately but permanently more effective and productive than on muck or sandy soils, and therefore more profitable in an economic sense.

and quality is so slow that very frequently it becomes uneconomic. A much better system is that of using rapidly available commercial chemical fertilizer to produce heavy and lush crops of green manure. In this process the chemical fertilizer (to meet the objections of the organic extremist) is converted into organic form through transformation rapidly into green manure in large quantity, which means that once plowed under, the elements incorporated in the original chemical fertilizer, converted into organic form, become rapidly and indeed immediately available in organic form to the crops which follow. This is a principle not understood or practiced sufficiently in this country.

One of our constant struggles at Malabar is to avoid becoming kidnaped by the organic extremists and even the cranks. I have observed earlier in this book and will probably observe again that at Malabar we are not now and never have been extremists or cranks. If we find the need for chemical fertilizer, largely to produce lush cover crops and green manures, we use chemical fertilizer in considerable quantities and with excellent effect. If we find that it is necessary to use an insecticide or a fungicide, we use them although we will always choose the organic forms of rotenone, nicotine, and pyrethrum or Bordeaux mixture, all of which are comparatively harmless to animals and humans. In *Out of the Earth* there are many pages devoted to the rather senseless battle between the chemical fertilizer "quick-buck" people and the organic extremists.

In all these operations we have now had nearly fourteen years of intimate and concentrated experience at Malabar and on this record we feel qualified to speak regarding many phases of living, productive soils which produce not only vegetables and crops notable for high and concentrated production per acre, but for quality in terms of flavor, vitamins, minerals, and general nutrition.

Although muck soils consist largely of nitrogen and carbon, the very base of organics, the case of sandy soils is quite different. If sandy soils are cultivated intensively, incessantly, and carelessly for a sufficient length of time, they will merely become layers of very nearly pure silica, soils in themselves very nearly sterile bacteriologically and utterly deficient and incapable of holding or maintaining either fertility or mineral balances. In the extreme extension of such a process, soils with a heavy sand content simply become lateral sand dunes and their capacity for maintaining any degree of balanced mineral fertility or moisture decreases. The first symptoms of disintegration of all fertility become evident in increasing incidence of disease and a feebleness of growth, which in turn make them subject to wholesale attack by insects. Such areas, subject to a continued exploitation which ignores organics and microbiology eventually reaches a condition closely resembling that of a desert.[6]

But a secondary evil developed. As the soils grew less and less capable of holding the chemical fertilizers where they belonged or of absorbing and maintaining moisture, the plants grown in such soils became steadily weaker and poorer in quality and speed of growth, and so subject to the attack of almost any disease and notably of those insects belonging to the beetle and aphid families. To correct this condition, increasingly vast amounts of poisonous dusts and sprays were employed at very great expense, not only for the purchase of the poisons themselves but for the airplanes and operators which applied these dusts and sprays in wholesale amounts. I have seen spinach grown in this particular operation so covered with DDT that the leaves were no longer green but brown. One evening while driving across one of the vast fields of vegetables, there appeared ahead of me a heavy whitish cloud that resembled fog. When I observed to one of the managers that we seemed to be a long way inland from the sea to have fogs, he replied, "Oh, that's not fog. The planes have just been dusting!"

In the cases of many of these poisons now used wholesale—although some of them are so virulent that the operator using them is cautioned to employ a gas mask—we know little or nothing regarding their eventual effects upon the human system. Years of research would be necessary to arrive at a definite answer, but such poisons are created, released, and

[6]I have seen one large "industrial" vegetable operation which, under intensive and exploitive operations in sandy soils over a period of years, is now faced with the serious problem of virtually reconstructing its soils, since in many areas they have been reduced to a texture and fertility resembling somewhat that of an ordinary beach. The land was originally chosen for vegetable operations because it was sandy and "easy to work." In successive years it was heavily cropped and as the organic material gradually disappeared with small or no effort at replenishment, the soils produced poorer and poorer crops both in yields and quality. To correct this, huge quantities of chemical fertilizer were poured on soils which rapidly became a kind of filter through which all liquids, including the chemical fertilizer in liquid form simply leached downward more and more rapidly each successive year to levels where it became unavailable to the plants. Finally, a great tonnage of this chemical fertilizer produced little or no effect and was even detected in considerable quantities in nearby streams and ponds far from the areas in which it had been applied. Rainfall and even great quantities of irrigation water also simply ran through the exhausted sandy soil, which became every year less and less capable of absorbing and maintaining the moisture necessary to all plant growth. Indeed, the more irrigation that was used, the more the basic evil condition was accelerated.

used wholesale overnight in vast quantities. The fact is that many a food producer, in all fields, is constantly looking for more and more virulent poisons which will do away more quickly and more effectively and more cheaply with the disease and pests for which his own exploiting operations are frequently largely responsible.

In this connection it is worth observation that at the present moment millions are being expended into research on cigarette smoking and its possible relation to the increase in lung cancer. Yet little or no research is being conducted into the possible relationship between the "smogs" of our great cities and the immense amounts of known poisonous fumes coming from trucks and automobiles, a carbon medium and agent which is much more obviously connected with cancer in all its forms, and can actually be used to create cancer in mice. Nor in all the research into cigarettes and lung cancer has much attention been given to a possible relationship based upon the virulent dusts and poisons with which most tobacco plants are constantly sprayed.[7]

The one corrective with sandy soils, just as it is with stiff and stubborn clay soils of much greater fundamental fertility, is the abundant incorporation into the soil of large quantities of organic material, which tends to create a sponge-like condition in which neither the natural mineral fertility nor the expensive chemical fertilizers leach away but are maintained at a level at which they may benefit the crops and make for high and quality production. In stiff clay soils the use of and incorporation of abundant organic materials simply tends to make for a workable and fine soil structure which, in the general pattern of growth, death and decay, sets in motion the bacteriological processes which produce fine and abundant

---

[7]On one recent occasion I arrived early on a foggy morning in one of the great railway stations in New York City. The cold foggy air from the streets above had settled into the lower levels of the station and when I stepped from the train I had a moment of actual terror when I first inhaled the accumulated fumes from the streets above and experienced difficulty in breathing until I reached the street level. Even there breathing conditions were miserable.

We already *know* that certain chemical combinations of carbon can actually *produce* cancer; We also *know* that the smogs and the ordinary fumes of the congested areas of our cities contain great quantities of carbon in various volatile chemical combinations, but in good American fashion this obvious factor is simply brushed aside in the lung cancer research in a hysterical concern over cigarettes. So far as I know, there has been no really comprehensive survey comparing the incidence of lung cancer among city dwellers as compared to those who live in areas of less polluted or wholly unpolluted air. At times one is tempted to believe that much of research and the work done in great industrial chemical laboratories is less concerned with human welfare, or even science, than with the stockholders' profits.

and healthy crops of high nutritional value, and to absorb and maintain moisture regardless of rainfall or atmospheric conditions or temperatures.

All these factors seem to me elementary and understandable to the mind of a child playing with a pile of sand in a kindergarten, yet it is astonishing how little they have been understood and how infrequently the vegetable grower or farmer has observed them or profited by them. It is, indeed, only lately that sufficient attention has been given to soil structure even in our large agricultural institutions. Sometimes this lack of observation and understanding arises from the greed of rapid production and high return per acre for a few years, until yields and quality begin to go downward, and the attacks of disease and insects steadily mount as the available fertility and the actual content of the minerals necessary to a decent nutrition for plants, animals, and people constantly declines. Presently the soil itself becomes unproductive, and the cost of sprays, dusts, and supplementary chemical fertilizers becomes economically prohibitive and the land goes back to the Indians.

This is a process not only observable in the past, but one which is in actual progress on many a farm today. In a way it parallels the extravagant use and frequently the vast waste of irrigation waters in the great Southwest, where many crops are over-irrigated and where I have seen many an artesian well pouring out a four-inch stream of precious unused underground water down the nearest ditch because nobody has bothered to shut off a valve and conserve what is the very life's blood of agriculture and vegetable growing over millions of acres in the arid areas.

As the fertility and the moisture-holding capacity of many of these sandy-land farms declines, the expense of vegetable production, in terms of fertilizers and dusts and sprays and labor, increases until the operation on a given area becomes unprofitable and frequently the land is abandoned entirely. In the case of some operations with which I am familiar—most of them on a large "industrial" basis—the alternative has already become that of abandoning the land or of spending thousands and perhaps hundreds of thousands in its restoration through the creation and application of vast amounts of organic material, usually in the form of green manures.[8]

[8]In semitropical and tropical countries the destruction through oxidation of organic materials in heavily worked and cultivated soils is much more rapid than in temperate zone areas, where for a large part of the year the temperatures are low or the soil remains frozen and the process of oxidation is totally arrested. One of the finest agricultural operations I know in the world is Cambui, a huge *fazenda* in Brazil owned by British capital. For years this great and productive plantation has annually laid aside a part of its land simply for

When I observed earlier that at Malabar we have lived "intimately" with soils and vegetables and vegetation in general for fourteen years, that was exactly what I meant, for our relationship with soils and the crops grown in them has been close indeed. In my own case this "intimacy" is of much longer standing. It means simply that one lives from day to day with soils and plants, observing the uncurling of each young leaf, the growth of the roots, which varies greatly among plants, the kind of soils in which a given plant flourishes, and of the kind in which it languishes and becomes sickly or dies. It means daily observation as to what insects attack certain plants, how they do it, and perhaps *why* they do it.

Back of all this, however, there must be something else … perhaps we should call it a kind of fire and enthusiasm … which makes this intimacy and observation in any field of activity the most important thing in the world. It is the same kind of fire which infuses the really good livestock man in relation to his pigs or his cattle. Without this fire all the education in the world is worth little, and all too frequently can lead merely to faulty deductions and errors. It makes of any task or livelihood not merely a question of working to feed one's self and one's family, or to draw down a

---

the "growing" of organic material to replenish the material destroyed by intensive operations in a semitropical area. This reserved area is planted to elephant grass, colonial grass, guandu or pigeon pea, and lab-lab beans, mostly imported into Brazil from Africa. These are very coarse fast-growing legumes and grasses, some of them attaining a height of ten to twelve feet. They grow rapidly and a huge tonnage per acre can be harvested through cutting them two or three times a year. They are run through an ordinary forage harvester and composted in huge piles before being returned to the land and crops in heavy proportions. One interesting factor is that, provided chemical fertilizers are used wisely, the grasses and legumes tend to improve the soils on which they are grown through the incorporation of nitrogen from the air and through the immense and coarse root systems of the plants. Therefore it is possible to grow these organic "supplements", with the addition of chemical fertilizers, on the poorer soils of the plantation and actually improve the soils, *while* harvesting at the same time vast quantities of organic material to be used elsewhere on the hard-worked fields. At Malabar-do-Brasil, which is largely a vegetable farm, and produces sometimes three or four crops a year on the same land, the same system of "growing" organic material is employed, but at Malabar-do-Brasil, a much smaller operation than Cambui, we are able to incorporate considerable quantities of the manure produced by subsidiary livestock enterprises including beef and dairy cattle, and pigs and chickens. While these various smaller enterprises are in themselves profitable, their principal purpose is to produce animal manures which can be mixed with the coarse fast-growing legumes and grass in the process of composting. Although in semitropical and tropical areas, organic material is destroyed rapidly under any system of row crops or frequent cultivation, it is also possible in the same areas to grow very rapidly a whole range of legumes and grasses which produce truly immense quantities either of forages for livestock or organic material and nitrogen through root and stalk … a production perhaps three or four times as great per acre as many grass and legume combinations that can be grown in the temperate zones.

salary and perhaps finally a pension: for the true research and experiment man there are no hours and no weather, and the plant, the pig, or the cow or what you will, becomes not only a goal in itself, but the *whole* goal.

This kind of devotion, passion, and interest can exist among the educated and the trained or it can exist among those who have educated themselves by observation, through books, and through contacts with others who share the same passionate tastes and interests and who therefore are among those that belong on the "inside." Some of the greatest scientific contributions have been made by practical workers, by inspired observers, by country doctors and farmers. Darwin, consumed by this fire and in the face of every obstacle and opposition, created an intellectual and scientific revolution. If one has a thorough formal education along with the fire, so much the better, but the education without the fire can be meaningless and frequently enough leads only to dust.

In all our experimental work in soils and plants and nutrition, at Malabar, I have been fortunate in some fields in finding the sort of man who not only possessed the fire, but the education. We have been most fortunate of all in the field of soils and vegetables and plants in general. The two young men, David Rimmer and Patrick Nutt, both English born and trained, who have been in charge of soils and vegetables for a long time now, have made very great contributions not only to our increasing knowledge of our soils but in general to the well-being of the whole farm and enterprise. In other fields we have encountered the same devotion, intelligence and interest notably in three young Swedes, Simon Bonnier, Olle Santesson, and Hasse Lindgren.

Under David and Patrick, soils which were infertile, unpromising, and unbalanced, have grown into soils of the greatest productivity, not only in quantity but in quality, and working from premises which seemed apparent to me even a generation ago, we have managed to grow more and more frequently plants which have shown themselves increasingly resistant to the attacks of both disease and of insects. Sometimes there have been violent disagreements among Patrick, David, and myself, frequently arising in the beginning from the fact that there are immense climatic differences between England, where the two young fellows were trained, and Ohio, where I have lived most of my life. Both Patrick and David had to become accustomed to the fact that here there were many more hours of sunlight a year than in England, that this sunlight was immensely more intense, and that under the heat and winds, the soils here, even the best of them, dried out more quickly than in England. Also that where frequently in England a too heavy mulch produced a menacing crop of snails and slugs, this did not occur in Ohio where the hard winter climate gave snails and slugs a difficult time. They had to learn too of the "wildness" of natural growth on good soils during the hot moist summers of our Ohio country. The sight of tomatoes, unstaked, unpruned, and growing like vigorous weeds in Ohio is always an astonishing sight to all North Europeans who come from countries where in order to ripen tomatoes in most areas they must be grown in greenhouses or hotbeds, or carefully pruned and staked near some sheltering wall that holds the warmth of the sun. It is astonishing to these Northern Europeans that one of our worst spring weeds are the wild tomato seedlings which spring up unwanted from the earth where tomatoes were grown the year before. And I have found again and again in different parts of the world that, although the fundamental rules of agriculture and horticulture are the same around the whole globe, small tricks in the growing of things will sometimes work in one climate and soil, but will not be profitable or effective in a different soil and climate.

Both Patrick and David had plenty of practical experience, as well as first-rate educations in agriculture and horticulture, and I myself had a lifetime of practical experience, plus vast amounts of scientific knowledge acquired from books, from scientists, and from pamphlets from every part of the world.

From the beginning we have had a common aim and interest ... to produce the finest of plants and vegetables, as nearly as possible free from the need for dusts and sprays, and to carry out and even perhaps create a program in soils and gardening which would produce not only the economic advantages of high production per acre, but the equally important ones of quality, both in flavor and in nutrition. Behind all this lay the overall aim of Malabar which might well be painted over our doorway together with the phrase, "Come and see for yourselves."

This aim and goal might best be stated thus—"To increase production and quality every year *while* reducing costs and steadily increasing the fertility of our soils."

This we have been able to do on the whole farm in general and in the experimental and research plots of the vegetable gardens in a really intensive fashion. We have even reached a production, under ideal conditions, of better than eight hundred bushels of potatoes per acre and more than thirty-five tons of tomatoes. In addition to this high rate of production (and the estimates and records are not ours but those of members of the National Vegetable Growers Association), both crops were clean and of the highest quality. The records of sprays and dusts were only a small fraction of those necessary even a few years earlier on the same soils, and certainly a very small fraction of those used in ordinary commercial production.

In the case of both crops a dust against blight and in the case of the tomatoes a spotty dust of rotenone wherever a tomato worm showed up, and a rotenone dust for flea beetles on the early potatoes comprised the whole program. In the case of potatoes I have not seen a potato beetle (the ordinary potato bug) in three years although there are many very large potato growers in our area, where potatoes have been a leading crop for nearly a century. In the earlier period of experimentation, potato beetles were common on all our potatoes, and heavy dusts were required to protect the plants. I do not pretend to know the whole of the answer to this experience which might indeed be called a phenomenon; I merely make the observation as the truth and as a fact. It parallels our experience with mosaic blight on celery, which I shall recount a little farther on.

In the case of these crops, as in many others, the indications were and almost always are that the healthier and more productive and faster-growing the plant, the less likely are the wholesale attacks of disease and insects. One interesting sidelight was the fact that the potatoes were grown in a soil which had and has a pH of seven to eight, almost universally considered a soil far too high in lime for growing potatoes. You will find a photograph of the freshly dug potatoes in this book; it speaks for itself.

The methods used in growing the tomatoes and potatoes were virtually identical (after all, they are first cousins). Both were grown in well-drained soils very high in organic material. Both, after the first cultivation, were mulched with alfalfa hay. Good moisture conditions were maintained on both crops through an overhead irrigation system and both benefited by the use of specialized soluble fertilizers through the irrigation system—a system which enabled us to pinpoint the fertilizer. Being in liquid form it was immediately available to the plants through the leaves as well as through the roots. Thus if the plants showed any signs of deficiency of any given major or minor element (and a skilled gardener can recognize signs of a given deficiency as easily as a doctor may diagnose measles from the symptoms), it could be virtually corrected overnight by applying the needed elements through the irrigation water.

Let us see what such a program adds up to in a practical sense. It means that after the first mulching, all weeding and cultivation became unnecessary, thus cutting out a vast amount of labor. The mulch likewise preserved the moisture and cut the necessity of watering or irrigation by as much as 60 percent or more. The mulch likewise kept the soil in which the roots were growing and in which the potatoes were forming constantly moist, cool, and loose, a factor of the greatest importance in the case of potatoes but of almost equal importance to the great majority of plants. Potatoes grow best when not too deeply planted, but they will suffer and develop disease and malformities when the soil about them becomes hot, dry and caked. They are very sensitive to overly high temperatures.

As any gardener knows, the tomato layers easily and will put out new roots from the stem and branches wherever there is coolness and moisture. If undisturbed, the tomato plant will literally cover the surface of the earth with a lacy network of fine roots which very possibly draw nutritive factors from the air itself as well as from the soil. Where the tomatoes at Malabar were mulched and left undisturbed, nothing injured or destroyed this fine network of tiny thread-like roots and so the health and vigor of the plant and therefore its powers of resistance to disease and the attack of some insects were unhampered. They were, in effect, growing naturally throughout the summer and the incidence of spots, cracks, or rot were extremely rare and on some varieties nonexistent. The yield was enormous and continued until the first killing frost. They did not suddenly give up during the heat of the summer, as carefully staked, pruned, and cultivated plants will frequently do.

At the end of the season the mulch leaves an immense amount of organic material to be turned back into the soil before a winter cover and green manure crop was seeded. Not only did the remaining tomatoes and the dead stalks go back into the soil, but with it the already half-decayed quantities of alfalfa hay used for mulch. On these two crops, as in the case of many others, we achieved the goal, "Greater production in quantity and quality each year, while reducing costs and constantly increasing the fertility of the soil."

This factor of temperature in soils surrounding the roots of plants, or as many commercial growers refer to it, of "soil climate," we have found to be of the greatest importance in the healthy and rapid growth of almost any plant, although there is clearly a considerable variation in their soil temperature demands, determined usually by the background and history of the given plant. In other words, the moisture and soil temperature requirements of the melon family (of semidesert origin) are very different from those of celery (essentially a marsh plant from the temperate zone).

In thermometer tests, we have been able to get readings as high as 120 degrees Fahrenheit and upward in open, well-cultivated, and unmulched soils, while it has been difficult to get a reading much above 70 in adjoining soils beneath a well-managed mulch. There is every evidence that the intensity of heat from the sun or even of burning winds with a high power of evaporation have little damaging effect upon the portion of the plant and leaves aboveground so long as the roots remain in moist and cool soils. Indeed, in many cases the hotter the sun and the higher the temperature, the more rapid and healthy the growth of a given plant, *provided* the proper cool and moist "soil climate" is created and maintained below ground.[9] While alfalfa hay is possibly the best of all mulches, because of its high protein-nitrogen-mineral content, virtually any material may be used for mulching. Cotton hulls or similar waste products are excellent and in the case of strawberries and certain other crops at

[9] Such a thing as a "dust mulch" has long been recognized among farmers and gardeners, and its efficacy is generally accepted as one means of conserving moisture. Actually, a "dust mulch" is created simply by constant cultivation so that a dry, dusty layer of soil an inch or two deep is maintained above the deeper soil. Such a mulch is by no means as effective as any mulch consisting of organic materials for a number of reasons. (1) The first rain destroys its efficacy and unless there is an abundance of organic material, creates a "caked condition over the whole garden or field. (2) To destroy this caking and to re-create the "dust mulch," a whole new cultivation is necessary, contributing each time added labor costs. (3) The fine thread-like surface roots of many plants, of which the tomato is a good example, cannot endure the extremely high temperatures and the lack of moisture existing in or just beneath the dust mulch. (4) Of course where overhead irrigation is employed, as in most home and commercial gardens, the "dust mulch" becomes impossible as it is destroyed with each irrigation and a condition of capillary attraction is immediately set up which simply draws moisture out of the soil to be evaporated by the sun and sucked up by the first hot wind. None of these conditions can exist where a mulch of organic material is employed.

Moreover, the "dust mulch" damages or destroys virtually all bacteriological and other life on the surface of the soil and to a depth sometimes of two or three inches. Beneath a mulch of organic material, beneficial bacteriological action and the action of fungi and molds is occurring under the best possible conditions and is actually accelerated, and greatly increases the availability to the plant both of the natural fertility and the chemical fertilizer.

Malabar, even sawdust is used effectively, provided the plants and the mulch itself are given extra nitrogen to counteract the drain from the soil of the nitrogen which is used by the bacteria in breaking down the cellulose structure. It is preferable to use hardwood sawdusts, as pine sawdusts tend to be acid and resinous. Paper and even aluminum foil are used by some gardeners as a mulch between rows of growing vegetables and flowers.

In fact many "organic" gardeners advocate a mulch of stones or very coarse gravel for fruit trees and shrubs, and at Malabar we have found this method thoroughly effective in maintaining soil climate, preserving moisture, and promoting vigorous, sturdy, and solid growth. In the case of one very old Norwegian spruce of great height, the laying of a terrace of flagstone around its roots brought about a complete rejuvenation of the tree so that a whole new growth, resembling a new young tree, occurred at the very top of the old tree. Its twin, only seventy-five feet away, remains a very old tree with every sign of age. The flagstone obviously maintained virtually all the moisture for the benefit of the tree and established a cool, almost *cold* soil climate highly beneficial to a tree coming originally from cold northern temperate areas.

Of course, the business of "mulching" field-grown row crops over large areas seems difficult and would apparently require much labor if done by hand, but it is not impossible to accomplish this mulching process mechanically. For example, potatoes may be heavily mulched immediately after planting and the stems will force their way through the mulch into the sunlight, as indeed many an old-fashioned farmer-gardener knows well. In the case of field-grown cabbage, tomatoes, or similar crops, the mulch can be applied immediately after fitting the soil and the plants set out in the mulch.[10]

[10]Mulching can be done mechanically and at the present time one or two agricultural machinery companies are working on a machine similar to a manure spreader which can put on quickly and easily mulches sufficiently heavy to do away with all necessity for weeding and cultivation. In addition to cutting immensely the labor costs, I am quite sure that the production would be much higher and the quality of the crop finer and faster-growing than in the case of ordinary weeding and cultivation. In such a process the plant roots suffer no damage whatever, and remain undisturbed to push their growth as far as they like in the loose, moist, cool, open soil. As one commercial tomato grower who had followed the mulch system observed, "Although no machine exists at the moment which will mechanically plant tomatoes in heavy mulch, I can afford to hire ten or twenty workers to set out the plants by hand in a day or two and still be way ahead of the game in the saving on cultivation and weeding and the big gains in quality and overall production." It is possible that within a year or two we shall have machines which will set out plants efficiently in heavy mulch.

While it would be difficult and perhaps impossible economically to mulch large fields of corn, one can arrive at a similar result in another fashion ... that of the "sheet composting" practiced at Malabar. In other words, the production of heavy quantities of organic material mixed into the earth to a depth of eight or ten inches produces a soil which not only absorbs all moisture like a sponge, but maintains it as well. A soil rich in organic material is in itself a kind of mulch protecting the lower soil from evaporation and from excessively high temperatures. At Malabar we have also found that on soils of sufficient fertility, the thick planting of corn, in soils heavy.with organic material, also served to cut evaporation and high, burning temperatures in the soil itself to a considerable depth. By keeping the earth in which the corn was growing completely shaded and by shutting out all wind, the moisture consumed by the extra corn plants was far outbalanced by the moisture that was previously lost in thinner plantings through penetrating hot sun and hot winds. In other words, corn planted thickly enough completely shades the ground and shuts out all hot winds, which are the greatest destroyers of soil moisture.

## MORE OF THE SAME

THE USE OF MULCH WHEREVER IT WAS POSSIBLE AND PRACTICAL IS BY NO means the whole answer to much of the resistance to disease and insects that we have been able to establish in nearly all garden crops at Malabar. There are countless other factors such as the deep tillage and sheet composting to a depth of at least eighteen to twenty inches which is now practiced at Malabar not only in the gardens but in the fields themselves. There is the question of proper mineral balances, not only in relation to the *general* soil condition but in relation to given crops and plants, whose appetites and demands vary greatly. For example, the demands for nitrogen in the case of celery, lettuce, spinach, and other leafy plants is very high, and in order to produce first-rate resistant crops these plants must be satisfied by giving them additional nitrogen fertilizer in one form or another. In contrast, overheavy and out-of-balance applications of nitrogen would be harmful to the production of such crops as tomatoes and potatoes, which on the other hand have an almost insatiable appetite for potash, which must be satisfied if the best results are to be obtained. Seed crops demand high amounts of phosphorus, and root crops need both potash and phosphorus in good quantity. And nearly all garden products, and field crops as well, demand a soil which carries a pH of at least five to six, which for the home gardener means simply that there should be sufficient calcium or lime.

I will return here to the case of potatoes grown at Malabar, with high yields and fine quality virtually free from disease, in a soil with a pH as high as seven to eight, which is usually considered far too high for the production of good potatoes, a crop generally regarded as preferring a slightly acid soil. We know that the pH in our potato plots was too high in lime, according to all general instructions on how to produce top yields and quality in the crop, yet the results belied all this information. We were able to produce one of the heaviest and finest crops of potatoes ever produced anywhere, on soils with an extremely high lime content, and when we began to look for the answer, which was certainly contradictory to most accepted rules, we came to the conclusion that the answer was one of *balance*.

In other words, a pH of seven to eight was not too high *provided* all the other elements existed upon the same high, abundant, and *available* level. Where the organic material was abundant, as well as the potash, phosphorus, and nitrogen, and in proper balance and ratio to the high lime content, the result was not only a good potato crop but a record one both in terms of quantity and quality. The element of the availability of these elements was equally important. Ordinary soil tests on many soils will frequently show large quantities of a given element, but do not reveal whether this element exists in a chemical combination which makes it highly available to the crop, or whether the soil texture is such that it will absorb and conserve the rainfall and moisture which are necessary to the availability and the activity of soil bacteria and other living elements vital in making both natural and chemical fertilizers available to the crops. There is much we still do not know regarding the microbiological relationship to availability.

In the case of the potatoes, there was in addition to the good and very high-level balance of mineral elements, a very nearly perfect soil structure, good drainage and soil climate, created in turn by the use of overhead irrigation, mulching, and the presence of abundant quantities of organic material ... all of which contributed unquestionably to the high degree of availability in relation to the mineral elements. In other words, the potatoes were grown under conditions as nearly perfect (according to available information and our own experimentation and research) as we could make them. The result was tremendous from every point of view.

These conditions are by no means impossible for the commercial grower to produce on a large scale. Many of the most prosperous potato growers aim each year to produce just such a combination of conditions. Our principal purpose in the case of potatoes, and with other garden-field

crops, has been to establish as far as possible what we believe to be the fact
that one can produce bigger and finer crops with a great reduction in the
incidence of attack by disease and insects by working through soil alone,
and by establishing *all* the conditions for a good, living, and productive
soil out of the information available and out of our own considerable
research and experimentation.

This seems to me to be a reasonable and intelligent method of procedure, although in many areas of agricultural research not only has it been
ignored, but its efficacy even disputed and assailed. In general, the modern academic movement has been to advocate a kind of hit-and-miss
agriculture and horticulture in which one attempts to cure the bad
results, the attack of disease and insects, or even the abuse of soils and
consequent low production, simply by using more and more fertilizer,
whether available or not, and more and more millions of pounds of violent poisons, the use of which *increases* annually as the texture, quality,
availability, and balance of great areas of food-producing land steadily
deteriorate.

In this question of balances we are aware at Malabar that there is an
important part played by the so-called minor elements, which in recent
years have become the subject of much discussion and controversy. The
minor elements, as all good commercial gardeners, fruit growers, and
farmers know, include such elements as iron, sulphur, copper, cobalt,
manganese, boron, molybdenum, and many others. Within the past few
years it has been established beyond scientific dispute that some of these
elements are absolutely *vital* to healthy and productive and resistant
growth in plants, trees, and shrubs, and some absolutely vital to the
metabolism, the well-being, health and even reproductive and disease-
resistant factors of animals and humans. Some of these elements have
been proved essential to plants, animals, and people alike. They are generally known as *minor* elements simply because they need exist in soils
only in small, even minute, quantities, although they can and do have a
very great effect upon quality and quantity production, and undoubtedly upon health and resistance.

It was not so long ago that quite an array of "authorities" maintained
loudly that these elements were nonessential and of no importance to the
production of healthy crops, animals, and people. This resistance—or
shall we say snap judgment and assertiveness—was broken down and dis-

credited with the discovery of vitamins and enzymes and the analysis of their molecular structure, to which many of these so-called minor elements are absolutely essential. In the case of plants, it was further destroyed by the actual pragmatic proof of the absolute necessity of adding to the soils, or to the plant structure through leaf spraying, certain of these elements in order to produce decent crops, or any crops at all. Perhaps the earliest and most striking work was done on citrus fruits and pecan trees grown on unbalanced soils in which some of these minor elements such as zinc, copper, and manganese were deficient or unavailable. While among the agricultural Brahmins, there is still some resistance to these ideas; there is none whatever among the commercial vegetable growers and the practical producers of fruit of all kinds. They *know*, for the most practical reasons which spell the difference sometimes between prosperity and economic ruin, that many of these minor elements are absolutely essential.

The opposition has now withdrawn to entrenched lines, in which it is asserted that it is not necessary to consider these elements since they are already present in most soils. While there is some truth in all of this, the question of the availability or nonavailability of these vital elements is not considered, and here again we return as always to the basic factor of soil *structure*, which plays so large a role in the availability not only of these minor elements but of the major elements as well. And the greatest fault of American agriculture today and the reason for the poor income and high costs of production for many a farmer, is merely that of *soil structure*, which is tied in so closely not only with the availability of natural fertility, but of chemical fertilizer as well, and above all with the absorption and conservation of rainfall and moisture (a factor that has become increasingly important in the present drought cycle). It is also related to the widespread deficiencies of nitrogen, which is perhaps the greatest factor in poor and sickly production and, to some degree, in the susceptibility of plants and crops to attack by disease and insects, and even to the decline of the protein, nutritional, and vitamin content of many food products.[1]

[1]Nothing illustrates this point of availability better than our experience with many field crops at Malabar growing under conditions or in areas where overhead irrigation was not possible. in periods of real heat and drought the plants would show signs, sometimes extreme ones, of deficiencies of a wide variety of elements in soils which were not actually deficient at all, or in which the elements were not even unavailable under average conditions of moisture and rainfall. It was only that under conditions of extreme heat and drought these elements in an almost totally dry soil were not available to the plants for the time being. The first heavy rainfall restored the availability and the signs of deficiencies disappeared from the crops. This condition has tended more and more to disappear

The fact remains, however, that the case for making certain that a very great range of these minor elements *are* in the soil (whether naturally or introduced artificially), and that these are readily available, is now well beyond any point of reasonable scientific controversy in a sound, productive, and healthy agriculture and horticulture. At Malabar we are fortunate in having soils in which virtually every one of the recognized minor elements, with the exception of iodine, is naturally present in reasonable amounts. Although it is difficult to gauge with complete accuracy the degree of availability of such elements in the minute quantities which are essential, we have every general physical evidence that as the topsoils were improved, deepened, or in some areas actually created, as organic content was increased and in general *a living, productive* soil has been produced, the availability of these elements has increased enormously.

This evidence is particularly clear and even striking in the case of alfalfa, which is one of the best and most accurate of soil-testing plants, not only because of its somewhat complex and in some respects almost voracious demands upon soil, but also because of its very deep-rooting habits and the undeniable capacity of its roots for breaking down and making available to its own growth and even that of plants which succeed it, soil mineral particles even though they existed originally in the form of large rocks and coarse gravels.

We have had a very long and intensive experience with this plant, which has been growing on the soils at Malabar since the very first year we took over its abandoned and *apparently* infertile fields. Alfalfa is very sensitive to deficiencies of the major elements, and especially to deficiencies of the minor elements, boron and manganese, and in its sensitivity it produces signs and symptoms in the growth of leaf and stem which are plain, clearly distinguishable, and marked evidence of given deficiencies. To be sure, if there is a serious deficiency of calcium, it will be difficult to grow any alfalfa at all; if there are shortages of potash or phosphorus the production will be limited both in quantity and quality; but the effect of these limitations is paralleled in the case of deficiencies or unavailability

---

at Malabar, even in hot and dry seasons, as the organic content of the soil and soil structure itself has been vastly improved. Indeed, one might almost say that these signs of "false unavailability" have continued to diminish in exact ratio to the improvement in soil structure and organic material, brought about largely through deep tillage and the process of sheet composting, which has been practiced for a number of years. Any farmer knows that chemical fertilizer or natural fertility becomes unavailable when there is insufficient moisture, and that no amount of top dressing with nitrogen will benefit a crop of corn if the moisture content and consequently the soil texture is poor.

of certain minor elements as well. If there is a lack or a serious deficiency of boron, the effect will be virtually the same as a deficiency of lime; it will be difficult or impossible to grow any alfalfa at all. The same facts are true of most other legumes in varying degrees. Yet there need be applied only as little as fifteen or twenty pounds of ordinary borax per acre to correct the condition and secure a good alfalfa, seeding. Manganese need exist in even smaller quantities in order to produce good alfalfa but its presence in the soil is equally indispensable. Molybdenum must be present in truly minute quantities for the healthy growth of any plant whatsoever, but in nearly all soils these almost infinitesimal quantities of molybdenum already exist.[2]

At Malabar there existed knolls and slopes of a gravelly nature subject in the past both to erosion and to the leaching action of heavy rainfall. Both conditions were created by an almost total deficiency of organic material; they provided the chronic "poor spots" in a field with which most farmers are familiar at one time or another. On these knolls, despite the well-drained character of the gravelly soils and the obviously high mineral content of their existing composition, our alfalfa seedlings remained thin or displayed the measles spots of boron deficiency, and in some places simply died out altogether. The application of borax presently corrected

[2]Over large areas in the red-soil South, it was found that even when all the major elements were present or applied in abundant quantities, it was still difficult to raise alfalfa or even good crops of the more common legumes, until it was discovered that the factor which prevented the growth of this "green gold" was the deficiency or absence of boron. Once corrected by the application of relatively small quantities of borax, alfalfa crops flourished to the immense economic gain of countless farmers in the old depleted cotton and tobacco areas. [I mention here three notable books or publications of the greatest value to farmers and gardeners in determining and correcting soil deficiencies, both of major and minor elements. One of these is an indispensable volume called *Hunger Signs in Crops* compiled by a group of the leading enlightened agronomists of the country, and published by the National Fertilizer Association of Washington D.C. The second is a booklet, very handsome and illustrated with many color plates, called *A Better Living from Your Soil*, and available from *Successful Farming* magazine, Meredith Publishing Company, Des Moines, Iowa. The third is a monthly publication called *Better Crops With Plant Food* published by the American Potash Institute, Washington D.C. The National Fertilizer Association has recently published a second book of indispensable value to those interested in good gardens and good lawns. It is called *The Care and Feeding of Garden Plants* and possesses the same high level of information as *Hunger Signs in Crops*. *Successful Farming* magazine has also issued recently a new and excellent booklet called *A Better Living From Your Livestock*.] Of course there is an increasing amount of information being published constantly in almost all agricultural and horticultural magazines. It is interesting to observe the increasing emphasis everywhere on soil structure and consequently of organic material as the very basis of any sound, productive, and permanent agriculture … a fact which good farmers and authorities in such countries as France, Belgium, Holland, and Denmark have recognized for generations.

the condition on some of these spots and we obtained fine and healthy seedings. The interesting fact is that the remaining poor spots, untreated with borax, tended to correct themselves as large quantities of green and barnyard manures were incorporated into the loose gravelly soil, and the whole of the soil's metabolism was corrected by the consequent increase in bacteria, minute life of all kinds, fungi and molds, and the retention of soil moisture. As the tendency to leaching was arrested by the sponge-like action of the organic material, we began to produce an abundance of the very finest alfalfa upon these same poor spots. The process was simply what might be described as a reactivation of the soils in these areas. A dead soil in which the mineral fertility and balance was actually very high was brought to life, and the minerals became available. The necessary amounts of boron were present, but simply unavailable because of poor soil texture and the "dead" condition of the soil itself.

The necessary elements were present all the time but were merely unavailable owing to the earlier practices of a miserable agriculture. Today these "poor spots" are no longer visible on any field of the thousand acres at Malabar. Because of the gravelly character of these glacial and alluvial deposits and the resulting good drainage, they produce not only our most vigorous alfalfa, but the finest in terms of animal nutrition.[3]

To be sure, all this discussion leads us still further into speculations regarding the deeper levels of the soil. In this respect, alfalfa leads us to almost unbelievable depths, for alfalfa roots have been traced at Colorado State College to a depth of 56 feet and at New Mexico College to a depth of 125. (These are not my figures but published official ones.) At Malabar we have traced the roots of our own alfalfa to a depth of twenty-five feet and then given up. One is tempted to ask at once what the roots are doing at such a depth and what are they finding there.

It is the custom at Malabar to run meadow mixtures of alfalfa, brome grass, and ladino clover for five, six, or seven years, taking annually three cuttings a year and on occasions four. Frequently, in order to maintain our rotation, we are forced to plow up such fields although they are still in a high state of productivity. We have at Malabar alfalfa plants which are thirteen years old with crown roots the size of parsnips which are still producing good crops of alfalfa.

---

[3]It is also true that by now the fields over the entire farm have become heavily inoculated with the bacteria necessary for the growth of good crops of alfalfa, sweet clover, or indeed of any leguminous plant. It is these bacteria which, forming nodules on the roots of the legumes, fix in the earth the nitrogen taken by the plant from the air.

Frequently we are asked by farmers, "How do you maintain heavily cut fields such as these for so long a time?" I am forced to say that very largely it is a matter of luck—that we have many fields in which the gravelly glacial-alluvial subsoils go down to a depth of thirty feet and more. Such fields have very nearly perfect drainage, which is essential to alfalfa, and an abundance and great variety of minerals, including dolomitic limestone which has been pulverized or ground up thousands of years ago by the action of glaciers and water. Our sources of subsoil lime are therefore of much higher availability than ordinary underlying solid limestone rock strata in so-called limestone soils. Thus we have in some of our soils great and varied reserves of mineral fertility which the alfalfa is able to utilize, and the maintenance of alfalfa on such soils is made much easier, despite heavy cutting, than on less hospitable and rich or shallower soils.

More and more there are indications that, with us at least, the maintenance of sturdy, abundant alfalfa over long periods is less a matter of giving it a "rest period", than of the natural high and readily available fertility of the subsoils. In other words, where there is high fertility, heavy cutting at reasonable intervals does not have any very perceptible effect upon the maintenance of the alfalfa. The plant is securing enough mineral nutrition deep down to maintain itself, save under conditions of heavy grazing, where no growth at all is possible in order to extend its root system and store up in the roots food reserve necessary to carry it through the winter.

The extraordinary thing is that only a short time ago these same soils, which produce such fine crops of alfalfa, and indeed of every crop, were regarded by some authorities and many old-time farmers as poor, unproductive, and scarcely worth restoring and farming. Indeed, when we first came to Malabar sixteen years ago, a field of alfalfa in the area was a curiosity, and the popular opinion was that it could not be grown in the area. Today alfalfa and alfalfa-ladino-grass mixtures exist on nearly every farm and have added millions to the farm income of northeastern Ohio, while serving at the same time to restore fertility to thousands of farms. Out of sixteen years of experience I think I prefer our gravelly and even our clay *hill* soils to any I know in the flatland country. We have no problems of drainage worthy of mention; we are able to work these soils at almost any time of the year without damage, and in the gravels we have reserves of mineral fertility which, if made available to plants by a proper agriculture, are virtually inexhaustible.

Much that we have observed, experienced, and accomplished at Malabar, in fields as well as in gardens and experimental plots, leads us to somewhat new conceptions regarding soil fertility and to doubts regarding all so-called "worn-out" land or farms. All soil consists of minerals in one form or another or there would be nothing there at all but air. Some soils have these minerals in good balance with few or no deficiencies, while others may be so ill-balanced and so deficient that to make them productive in any true sense would be economically inadvisable or impractical under the present ratios between population and food needs. More and more it seems to us that the problem of fertility, on soils of good mineral balance is very largely one of availability of the natural fertility plus the wise use of chemical fertilizers if record production is the goal. Much of this availability is dependent upon the factors which have been emphasized perhaps to the point of boredom in this chapter, but to us at Malabar they seem to be absolutely vital factors which turn up everywhere in all our experience, observation, and experimentation. These factors are soil structure, organic material, the absorption and conservation of all rainfall in areas of reasonable fall, the constant incorporation of large quantities of organic material, and finally the use from time to time of the deep-rooted grasses and legumes which have been discussed at length already in *Out of the Earth* and in another chapter of this book.

In other words, it seems to us more and more evident that a sound horticulture or agriculture is not based merely upon cultivating, fertilizing, and exploiting the top seven or eight inches of the soil, but upon undertaking a much deeper agriculture which tends in time to employ and utilize the fertility down to great depths. Modern machinery, power-tillage machinery, and deep tillage, together with the deep-rooted grasses and legumes, tend more and more to make this use of deep fertility possible.

It should be observed that this deep tillage and the use of deep-rooted grasses and legumes in rotations are especially valuable to the farmer or gardener with sandy, gravelly, and light soils, where heavy leaching, both of natural fertility and chemical fertilizers, occurs. Fertility leached to depths below which the root systems of ordinary crops do not extend, is picked up and utilized by the deep-rooted plants which return the fertility elements to the surface, either in the form of highly nutritious forage crops or as organic fertility in the form of roots and green manures.

To some degree, this is a wholly new conception of agriculture and horticulture, especially to those farmers and authorities still entrenched in the conviction that a sound and permanent agriculture is a matter merely of the exploitation and shallow cultivation of a few inches of soil near the surface.

It should not be overlooked that the top shallow level of soil is constantly subject to three heavy drains, (1) by the demands of the plants themselves, (2) by the constant destruction by heavy cultivation of organic material, and consequently of soil structure and moisture conservation, and finally (3) by the downward leaching action of heavy rainfall, especially on light or sandy soils. The conception of agriculture *in depth* tends to correct to some degree all these drains and to permit the roots of plants to feed where they will, and in the case of deep-rooted, green manure crops, to bring up from considerable depths valuable minerals in the organic form of roots, leaves, and stems, which in time come to be incorporated in the top few inches as highly available organic material and natural fertilizer.

The application in sprayed form of certain elements for absorption through the leaves of plants, and especially of trees and shrubs, is not new, but the application of specialized *whole* soluble fertilizer formulas through irrigation of one kind or another is comparatively recent, and its great advantages, economic, nutritional, and as a preventive to disease and insect attack, are only beginning to have a much broader understanding and application. For many years fruits and nut growers have been correcting deficiencies of minor elements such as boron, copper, manganese, zinc, and others by spraying these in liquid form upon the foliage, which is able to absorb them into the growing plant structure.

Five years ago at Malabar we began working with soluble fertilizer formulas which included some twelve minor elements on garden and field crops. Besides David Rimmer and myself, three other persons were involved in the production of the fertilizer formulas and in running down the results. One of these was Cliff Snyder, president of the Sunnyhill Coal Company, an engineer and inventor and a man with an especially lively, inquiring, and open mind. A second was Dr. Jonathan Forman, editor of the *Ohio State Medical Journal*, former president of the Society of Allergists and the Friends of the Land, and a specialist in nutrition and allergies. The third was Dr. J.D. Rebbeck of Pittsburgh, a surgeon and nutritionist. Much intensive experimental and research work was also done by the Battelle Metallurgical Institute of Columbus, Ohio, one of the most generously endowed and progressive research agencies in the world. Ollie Fink, executive secretary of the Friends of the Land, also made notable contributions of energy, work, and suggestion. Our own interest in Malabar was concerned both in quantity and quality of production *and* the nutrition of plants, animals, and people. It was an interest based, as always at Malabar, on the soil itself as a fundamental corrective of many human ills. As always, we were also concerned with the economic aspects as they affected the income of the average farmer and market gardener, and the question of feeding properly a world and a nation whose populations increase at a terrifying rate.

At Malabar we are well equipped with overhead irrigation in its most modern forms, and the problem was one of providing a simple and practical mixer which could dissolve quickly and under accurate measurement the fertilizer as it entered the irrigation water. Cliff Snyder quickly designed a simple and practical apparatus, working upon principles which were applicable to a garden hose, an ordinary lawn sprinkler, and large overhead irrigation systems up to the six-inch high-pressure variety. The final question was one of a balanced formula, which was worked out after consultations among those interested and with the aid of certain experts and the advanced information available. The last element was not very great since comparatively little work had been done in the whole field. The final general purpose formula involved for lawns, gardens, and general field crops was as follows.

TRACE ELEMENTS

Analysis of Fertileze (sulfate)

| | | |
|---|---|---|
| *Sulphur* from the sulphates | 3.19% as sulphate | 9.55% |
| *Magnesium* from the sulphate | 1.28% as oxide | 2.12% |
| *Calcium* from the phosphate | 1.48% as oxide | 2.07% |
| *Manganese* from the sulphate | 0.39% as oxide | 0.50% |
| *Iron* from the divalent sulphate | 0.14% at oxide | 0.20% |
| *Zinc* from the sulphate | 0.08% as oxide | 0.10% |
| *Copper* from the divalent sulphate | 0.05% as oxide | 0.07% |
| *Boron* from Borax | 0.01% as borate | 0.05% |
| *Cobalt* from the divalent sulphate | 0.01% as oxide | 0.02% |

| | | |
|---|---|---|
| *Iodine* from potassium iodide | 0.02% as iodide | 0.02% |
| *Molybdenum* from sodium molybdate | 0.01% as molybdate | 0.01% |

MAJOR ELEMENTS

| | |
|---|---|
| 20% | *Nitrogen* |
| 10% (pentoxide) | *Phosphorus* |
| 20% (oxide) | *Potassium* |

The product known as "Fertileze" also exists in a variety of other formulas, designed to meet the special needs of special crops and soils. These formulas include the same minor elements, with variations in the amounts of major elements. This soluble fertilizer, which is employed at Malabar, where much of the research and experimentation were carried out, is manufactured together with the flow-mixing equipment, by Nutritional Concentrates Inc., New Lexington, Ohio.

From the very beginning, the whole system and operation proved its efficacy in many ways as a means of producing high production, more rapid growth, and high quality both in nutrition and flavor. As the experiment progressed, other formulas were established which were especially adapted to certain ranges of crops with the elemental soil balances and the specialized demands of given crops under consideration. In other words, a formula high in potash was developed for potatoes and tomatoes, and another high in nitrogen for such crops as celery, lettuce, and sweet corn. The experimentation with specialized formulas still continues.

The advantages of liquid fertilizer are apparent to any experienced gardener. Liquid fertilizer is immediately available to the plant and there is no necessity to await rainfall or to irrigate heavily to make a dry fertilizer available quickly to the plant or the crop. The capacity of plants to absorb nutrients through the leaves, in the case of certain elements and minerals, more rapidly than the roots are able to absorb and utilize the same elements made it possible to feed very quickly almost any plant with exactly the formula and combination of elements it required for the finest production and quality. Moreover, it made it possible to "pinpoint" the fertilizer program. In other words, if a plant at any time showed a sign of a deficiency of any given element, it was possible to correct the deficiency almost immediately by the use of soluble fertilizer applied through irrigation. One can virtually see the effects of an application of soluble nitrogen the morning after it has been applied.

This soluble fertilizer program, combined with overhead irrigation, made very nearly possible the dream of every gardener and farmer (1) to make himself completely independent of the weather and of droughts, (2) to be assured at all times, regardless of conditions, of a nearly perfect bumper crop, (3) to correct any deficiencies quickly and almost immediately *during* the growth period of the plant and without waiting for rain or for another year. To put it simply, the system enables the grower to be certain always of a very nearly perfect crop of vigorous growth and fine quality, both nutritionally and in tenderness, flavor, and related qualities.

On the economic side, the process and system had other great advantages. It cut tremendously the costs of expensive special machinery and of labor in side dressing or making fertilizer applications *during* the growth of the plant. We discovered almost at once that, especially on good soils, the intelligent use of soluble fertilizers produced a very rapid but sound growth in vegetables. In early-season crops this greatly increased the income for the market gardener by producing crops earlier than many competitors, when the price was still high. In other words, we found that through the use of these soluble fertilizers in irrigation we were able virtually to wipe out the two to three weeks' difference in season between vegetable crops in our northern Ohio area and those along the Ohio River and the Middle Southern states.[4]

After two or three years of experimenting, it became clear that the advantages of the soluble fertilizer—varied formula—irrigation system were not only very great in relation to production, but in many other ways as well. Largely speaking, the more rapid the growth of the vegetable, the greater its tenderness and flavor. A natural consequence is a higher vitamin and mineral content. If we could achieve these things *plus* a higher resistance to disease and insects which would greatly reduce the need for poisonous dusts and sprays, we should be arriving somewhere near the goals which we had set long before in our general soils and crop program.

---

[4]In 1954 there occurred in our area a curious situation in which local buyers were unable to purchase beets, even from the shippers and wholesalers. The beet crop from the Deep South was exhausted, and the early beets of local production had not yet appeared in the markets of northern Ohio. At Malabar we had an abundance of early beets of fine quality, and retailers actually purchased beets from us at retail prices in order to provide them, with or without a markup, for their city customers. This despite, the fact that with our new roadside market we were actually competitors, although in a very small way. The beets we sold were actually grown in eight weeks from seed.

Actually, this is what happened. We did find a steady decline in the need for dusts and sprays. Let me repeat that at Malabar we are not cranks, and use dusts and sprays when necessary, making only the choice of organic and vegetable poisons rather than high-powered inorganic chemicals of which very little is known beyond their deadly effect upon the insects themselves and possibly in the long run upon animals, wild and domestic, and upon humans.

Today many crops grown at Malabar receive no dusts and sprays whatsoever throughout their period of growth. This is true of beets, carrots, celery, onions, parsley, and many other crops and many flowers. On the whole, we are using today less than 5 percent of the dusts and sprays which were necessary fifteen years ago in the same soils and general areas. In 1954 we began to move into experimenting with flowers as well as vegetables, and among notable results were the roses which, mulched and growing in fine, balanced soils and treated to soluble fertilizer through overhead irrigation, were untouched by disease or insects save for a mild attack of black spot on one variety of rose which appeared to be especially susceptible; and this despite the fact that this was the first season for the roses so they had not yet established strong and vigorous root systems.

Potatoes and tomatoes, both of which are dusted and sprayed almost constantly by the average grower, received only one or two mild applications of rotenone, or pyrethrum and Bordeaux mixture; string beans, usually devoured voraciously by the Mexican bean beetle, which is ubiquitous in the U.S., received in some cases only one dusting with rotenone and in the case of the earlier crops none at all. In some of these cases many gardeners would not have regarded the dusts and sprays as actually necessary, but both David and Pat are meticulous perfectionists and take no chances on producing a very nearly perfect product.

With the passing of time, two other elements have had to be considered in the efforts toward the gradual elimination of disease and insect attack, and consequently of the necessity for poisonous dusts and sprays. One, in the case of insects, is the bird population, which has greatly increased at Malabar Farm with the introduction of multiflora rose hedges. The hedges, thick, thorny, and impenetrable to bird enemies of all kinds, are especially favored by the birds of the thrush family, as well as by pheasant and quail. And the thrush family is voracious when it comes to insects, a young thrush consuming sometimes twice its weight in insects in a single day. The vegetable experiment and research plots at Malabar are nearly all enclosed by multiflora hedges which have attained a thick, entangled growth in which one can find a bird's nest on an average of every ten feet. It is certain ... and verified by observation ... that the birds go to work in the gardens searching for insects in the early morning before any worker in the gardens disturbs them, and, while it would be very nearly impossible to make a completely accurate record of the number and amount of insects consumed, it is certain that the birds carry off vast numbers of insect enemies.

The armies of birds following the mowing machine or hayrake, consuming insects in the open fields, have been described elsewhere in this book. These armies exist because of the policy maintained from the very beginning at Malabar of creating ideal conditions for wildlife by permitting fencerows to grow up, by the establishment of multiflora hedges, and even "islands" of undergrowth or multiflora and pines in large open fields. Even the unpopular house sparrow makes a contribution in keeping down insect populations, a contribution far beyond that for which he is given credit and which may have developed with the passing of the horse and the horse manure in which he once found an easy field for foraging. In France I have seen the house sparrow in our gardens stripping every green aphid from the roses.

The hedges, the birds, are merely a part of a general pattern based upon natural checks and balances in the operations of nature and have contributed enormously to the material welfare of Malabar Farm, as well as to the pleasure and beauty contributed by the hedges and by the presence and increase of bird life of every sort. In a purely scientific sense, the presence of the birds and their voracious appetites have somewhat upset our observations and calculations with regard to the establishment of resistance in plants to insect attack, for we cannot be sure exactly of the degree to which we and the soils deserve responsibility for increasing resistance and how much of the credit should be given to the birds, especially since the degree of unpalatability of certain insects to certain birds is difficult to determine with complete accuracy.

The other element we have come increasingly to consider is related to disease resistance, and it is one that we have long suspected but which we could neither pin down nor analyze because we lacked the time and labor,

Pleasant Valley

The Bromfield's "Big House" at Malabar Farm

Young Louis Bromfield at an Ohio farm

Ambulance driver in France during World War I

The "Jungle Room" at the Bromfield home in Senlis, France (at the left are Gertrude Stein & Alice B. Toklas)

The Presbytère at 2 Rue Saint-Étienne, Senlis, France

Louis Bromfield with his siamese cat Sita, at work on the novel *The Farm*

Louis Bromfield

on the beach with writers Edna Ferber, F. Scott and Zelda Fitzgerald

on a tiger hunt in India with Maharajah

Louis Bromfield at his beloved Malabar Farm

Photographs courtesy of: • Malabar Farm State Park • The Ohio State University Library • the Ohio Historical Society • and the Louis Bromfield family. Used with their permission.

Louis Bromfield in conversation with First Lady Eleanor Roosevelt

... and with fellow farmers

with farm manager Max Drake

with daughter Ellen Geld in Brazil at *Fazenda Malabar-do-Brasil*

"In this very great phrase 'Reverence for Life,' I too found what I had sought for so long."
—Louis Bromfield

the money and equipment. As in many other fields we were content to make an observation or suggestion and leave the actual detailed research to the specialists and the laboratories. This element was the relation of antibiotics to the health of plants, and the actual availability of antibiotics in soils high in organic materials and consequently in the fungi and molds from which all antibiotics are derived. In other words, we suspected that in *living* soils with a high organic content in which the cycle of birth, growth, death, decay, and rebirth was in healthy operation, antibiotics were constantly being created which might be available to the plants and so prevent or check the attack upon the plants of many kinds of plant diseases.

It was only in 1954 that the department of agriculture at Rutgers University, which has made such immense contributions to agriculture and the advance of nutrition in plants, animals, and people, announced the discovery that antibiotics sprayed upon plants and trees were quite as effective in preventing or checking disease as they were in accomplishing the same ends with animals and people. Diseases of certain vegetables and shrubs and trees, hitherto regarded as unpreventable or incurable, were easily prevented or cured by spraying them with almost minute quantities of antibiotics. Perhaps the most notable example was the case of fire blight, which attacks trees and sometimes whole orchards of apples and pears and is fatal to a considerable range of ornamental shrubs and trees. Few diseases could be more disastrous in an economic sense, for it not only damages trees and orchards and cuts production enormously, but in some instances it destroys whole orchards. It was found that this blight, which in the past could be neither prevented nor checked, was simply abolished altogether when a spray of antibiotic was used. Similar results were found in the case of many other bacterial blights affecting a wide range of plants.

The first reports asserted that there was no evidence that plants, shrubs, or trees could assimilate the actual antibiotics directly from the soils, but more recent research tends to show that this is possible and even probable. If this factor is once established, it leads backward into many of the practices we have long used and studied at Malabar: (1) the importance of large quantities of organic material, not merely dead and inert, but in a constant process of decay and so constantly producing fungi, molds and antibiotics, (2) the use of mulches, which maintain temperatures and steady moisture content encouraging to fungi and molds, (3) the fact that through the use of mulch a process of constant birth, growth, death, decay, and rebirth is established and encouraged, (4) that beneath the established mulches, the fine hairlike roots of many plants, which are responsible for any mycorrhizal action and are undoubtedly the means by which plants absorb many nutritive elements from the soil, remain undisturbed and uninjured either by high temperatures or by constant damage from cultivation, and so can carry out their functions to the fullest possible extent in absorbing nutrition, and even quite possibly the antibiotic substances produced by the fungi and molds at work in the soil and in the very damp, cool, decaying mulch material with which they are in direct contact.[5]

It is notable in this respect that the incidence of fire blight and other diseases were and are markedly less in orchards planted on new soils or in orchards which are properly treated to the benefits of legume plantings and mulches.[6]

As we have progressed year by year with the experimental work, it has become increasingly evident that there is something seriously wrong with the old-fashioned theories that the lush, heavy, fast-growing plants are ones which insects and disease prefer. In every case with every plant we have observed that this is not true. The plants they prefer and attack are the sickly plants, those suffering from deficiencies of this element or that one, or plants suffering from soils that are deficient in organic material or are too dry or too poorly drained. Any really observant gardener knows that where there is a variation in the health and vigor of plants in a given row or plot, it is always the sickly or weakened plant that is first attacked

---

[5]Recently impressive progress has been made in unraveling the extraordinary mysteries of photosynthesis, the process by which the plant, provided with sufficient mineral nutrition, produces more than 95 percent of its growth out of sunlight, air, and water. It is not improbable that further discoveries in this field will throw more light upon the whole problem of the resistance of plants to disease and insect attack.

[6]One of the finest and most prosperous farmers and fruit growers in the nation is Cosmos Bluebaugh, a Master Farmer of Ohio, who lives not far from us. His orchards are notable for productivity, quality, and freedom from disease, and for years he has raised alfalfa "not," as he puts it, "to feed my cattle but my orchards." He uses the alfalfa as a mulch and so establishes about his trees a continuing and active cycle of birth, growth, death, decay, and rebirth in which the fine roots of his fruit trees participate. It is a practice which unquestionably is responsible not only for the cutting down of disease attack, but also contributes to the quantity and quality of production and the long, vigorous life of the trees. Mr. Bluebaugh was never especially concerned with the disputes and wrangling over the efficacy of his practice. He simply used his instinct and his brains in the beginning, found that the system *worked*, and simply continued it year after year, to his great profit and the establishment of his reputation as one of the nation's best farmers.

by disease and in many cases by insects. The truth is, I think, that there has not been enough close, meticulous, and really scientific observation of the factors involved, that *all* the factors have not been considered and taken together, and that in too many cases the "specialist" and the "authority" working down his single narrow alley simply runs for the dust or spray at the very first sign of attack.

I have also heard it stated by some "authorities" that the mineral content, the vitamin content, and the general nutritional qualities of given plants are the same, regardless of whether they are grown on rich, living, balanced soils or on poor, unbalanced, deficient ones. If the evidence of the eye and taste were not simply enough, tests have shown this is not true. Observe the enormous difference in taste, consistency, and quality of a carrot that has been grown rapidly in a good soil in six or seven weeks to that of a carrot which has grown painfully and slowly in poor soil over a period of two or three months. The one is a delicious, sweet, tender, crisp, and nutritious vegetable; the other woody, knotty, tasteless, and hard. The poor color of the carrot from poor soil and the cellulose–carbon content alone tell the story.

Beans of any kind, but particularly the garden varieties such as string beans, French horticultural beans, lima beans, etc., are especially susceptible to disease and insect attack when grown under poor conditions, and are therefore one of the very best mediums for experimentation, testing, and research in the whole field of plant resistance. We have found that if any of the garden beans are given as perfect conditions as possible, the danger of attack by disease certainly and of insects partly is reduced and occasionally nonexistent. But give them too much or too little moisture, too much shade or too many weeds in competition with them, and both disease and insects will appear in quantity.

During the war when labor was scarce, rows or plots of beans at Malabar sometimes suffered from heavy weed competition and immediately became subject to attack. Once the weeds were removed and the plants had a chance to recover their vigor of growth, the attacks diminished or disappeared altogether. In 1954 we attempted the experiment of growing French horticultural and Kentucky wonder beans together with silage corn in the hope that they would climb the stems of the corn, do away with the trouble of staking them and still produce a good crop. The beans started off well and looked vigorous and healthy for the first three or four weeks; but we reckoned without the vigor of the corn in the same rich soils in which the beans were growing. The corn grew rapidly and vigorously, eventually reaching a height of ten to twelve feet, and as it grew it presently shut out virtually all sunlight from the climbing beans. Immediately the beans became harassed alike by disease and insects, very possibly because they had lost through lack of sunlight some of the elements vital to the whole process of photosynthesis and consequently to their vigor, health, and powers of resistance. The striking thing was that on the edges of the plot, where the beans on the outermost row and at the ends of the field had access to abundant sunlight, the attacks were almost imperceptible and certainly not worth the trouble of dusting or spraying.

One other striking factor has been observed in the Malabar plots of bush beans regularly over a period of years. While these beans, especially the later crops, are sometimes attacked by bean beetles—and bean beetles are nearly always present although in negligible quantities during the growing period—the full attack does not come until the crop has been harvested and the bean plant has begun to fade and die. Then the bean beetles appear in great numbers to attack and virtually destroy the sickly, dying plant. However, in the case of an occasional plant, there suddenly appears a kind of second flowering, consisting of a shoot or two of new and vigorous growth appearing from the very midst of the dying plant. Although the rest of the plant may be covered with beetles, the insects rarely if ever touch the new and vigorous growth ... surely an indication of the true function of the beetle to destroy sick, dead, or dying plants and reduce them to soil. The indication is that a really healthy and vigorous plant contains elements which make it distasteful to the beetle. This is not a casual or occasional occurrence; it is observed year after year. The infestation of bean beetles *after* the crop has been harvested is of no importance to the grower of string beans as, with us at least, the dying plants are quickly turned into the soil and converted into humus, and the beetles themselves actually serve as fertilizer. The same might *not* be true for the bean grower who is producing ripened, dried shell beans for the market, although the infestation would make no difference to his crop yields after the beans had formed in the pod and begun to ripen.

Perhaps even more striking is the behavior of the common so-called "squash bug", which appears in plantations of squash or pumpkins late in the season after the fruits have begun to ripen. They will *never* attack the green, vigorous, growing tip of the vine but only the injured, old, and withering leaves. Bend over and injure a healthy leaf and the moment the signs of injury become evident, the "squash bugs" will congregate on it in such numbers that there is "standing room only." This is an experiment which can be made by any amateur gardener if he has doubts.

It has long been the practice of experienced market gardeners to avoid growing the same crop on the same soil several years or even two years in succession. The purpose behind this is to avoid the risk of putting the new crop into soils heavily infected by the previous crop with the bacteria or spores of disease. At Malabar it has been interesting to observe over a period the efficacy or nonefficacy of this undoubtedly wise practice. In many cases we have grown the same crops on the same plots for many years, and in the area of our experimental plots, which is limited, there is always a likelihood that the bacteria and spores of disease are lurking somewhere in the immediate vicinity.

In 1954 the tomato crop provided good tests because part of the crop was raised on new land, part on land where tomatoes had grown the previous year, and part on land which was being used for tomatoes a third year. The only disease of tomatoes which has ever given us any trouble at all has been wilt, and it is undoubtedly true that to a small extent at Malabar, as elsewhere, the likelihood of attack by wilt is somewhat increased where the crop is grown on the same ground two or more times in succession; however, even under such circumstances, the attacks have never been serious. In 1954 it was not serious—indeed far below that observed in almost any tomato producer's fields where the crop was being grown in soils of poor balance and texture—one dust sufficed to stave off the attack save in the case of one small area which almost accidentally provided us with a striking example of the susceptibility of sickly plants to attack by disease.

In the tomato plots there existed a small area, not more than twenty-five feet in diameter, which, under conditions of heavy rainfall or over-irrigation, becomes poorly drained. The cause is an underlying layer of blue shale, isolated there at some remote period of glaciation and alluvial action. In the summer of 1954, during the hot dry weather in August, this plot had been given a heavy irrigation, and within twelve hours there followed a cloudburst downpour of rain. As a result the poorly drained area, in the very midst of the tomato plots, became literally saturated with

water; the water remaining in pools on the surface. Gradually the standing water disappeared, but at the same time, the tomato plants within the
poorly drained area broke out in a fiery rash of wilt which virtually
destroyed the plants. One of the most striking facts was that the virulence
of the attack declined gradually in circles away from the central point
where the worst saturation occurred. In other words, the plants which
had been virtually under water were practically destroyed by the disease,
those a little farther from the central point, growing in soil which suffered
somewhat less, were less affected and so on outward until in the well-
drained areas of the plot where the flood of water had been absorbed by
good drainage deep into the earth, there was little or no attack by wilt.
Nor did the wilt spread later to the adjoining well-drained healthy areas.

Clearly the evidence was that where the plants had been weakened by
poor drainage conditions *within* the soil, they became sickly and deprived
of the elements of rapid, vigorous growth and resistance. In this case no
special factor bred into the plant was involved in their resistance or susceptibility to attack; rows of four different varieties, some advertised as
wilt-resistant, some not, ran across the poorly drained area. The result
was the same on plants of all four varieties. I think it should be noted here
that the excellent plant breeders, who have spent years in breeding special
varieties of crops which exhibit resistance to disease, should not be held
responsible or their theories questioned when the bad or careless gardener or farmer finds a lack of resistance in these special varieties when he
grows them in poor soils and under poor conditions. The elements of
good soil and good soil practices still remain fundamental in the field of
vigor and resistance.

The story of celery at Malabar to which I referred earlier in the chapter is one of the most interesting. I have told it before in *Out of the Earth*
in much greater detail than I mean to employ here. We have grown celery
at Malabar for fifteen years; some of it year after year on the same ground.
In the beginning, for five years or more, the celery was badly infected by
blight and, for the sake of experiment, the leaves affected by blight were
year after year thrown on the ground and worked into it so that the actual infection was as great as it could possibly be. During that five-year period the soils were given an abundance of barnyard and green manures and
the soil dressed with applications (1) of Es-Min-el, a product of basic slag
produced in Alabama and high in trace element content and (2) of Sea

Soil, the residual by-product from the Dow Chemical Company operations in Texas in reclaiming of bromine, sodium, chlorine, and magnesium from sea water.

As the soil treatment continued and the minor elements eventually became highly available, and the soil texture and drainage greatly improved, the incidence of blight even in the badly infected soils began to decline at a noticeable rate. Each year the celery was cleaner. It was David, who by extra applications of nitrogen, succeeded in knocking out the blight altogether.

Celery has a tremendous need and appetite for nitrogen and for water, but even abundant irrigation and the use of barnyard manures and high-nitrogen fertilizer formulas at Malabar did not, apparently, satisfy its hunger for this basic element. The additional applications of nitrogen fertilizers produced the strength and vigor which finally knocked out the disease entirely. Although the average commercial grower sprays or dusts twice a week or more to control blight in celery, at Malabar we have for years now been growing absolutely blight-free celery of the finest quality without any dusts or sprays whatever, even on soils infected deliberately for years with the disease.

This was a notable case where the balance of elements for a special, given crop was wrong, and the proportion of one element (in this case nitrogen) not nearly large enough. A similar susceptibility to blight and disease occurs in tomatoes and potatoes where there is a similar deficiency of the element potash. The really expert gardener and farmer will not overlook this question of the demands of special plants for large quantities of given and special elements. A general fertilizer can be effective but not within miles of the fertilizer formula designed especially for the given demands of a given plant or crop. It is in this field that the system of soluble fertilizer in specified formulas for a given crop has worked brilliantly at Malabar in permitting us to pinpoint the treatment of special demands of special crops, and thus to produce something close to the optimum in production, quality, and resistance.

Not only do given plants have special given appetites for special given elements; they also have a decided special preference for the moisture contents of soils, a factor which many a home gardener and sometimes the commercial gardener overlooks altogether. Celery, in reality a marsh plant, can utilize an immense amount of water, provided the drainage is reasonably good; so can corn and some grasses. On the other hand, such vegetables and fruits as melons, squash, cucumbers, and others which are of desert or semidesert origin can be quickly drowned out or made highly susceptible to disease by over-irrigation or too much rainfall.

One of the initial difficulties at Malabar was the segregation of plants and crops so that they could be provided with the varying degrees of moisture which they needed and demanded. In other words, if cantaloupe was grown side by side with celery, so that the overhead irrigation system provided both with the same amounts of water, disaster was foredoomed for one or the other of the crops, or mediocrity of quality and production and susceptibility to disease for both of them. If sufficient irrigation was provided to grow the finest, most vigorous celery, the cantaloupe was certain to be drowned out and ruined. If, on the other hand, irrigation was limited to the moisture demands of the cantaloupe, the celery would be woody, wretched, and with poor resistance to disease. If a compromise was attempted, the special demands of neither plant would be satisfied, and poor quality and production and lack of resistance would be inevitable.

The problem of matching the irrigation to the needs and demands of the plant at Malabar was solved by greatly expanding the experimental and production plot areas and segregating the crops according to their needs and demands. With this change many difficulties were corrected and much disease and insect attack was eliminated at the very source. These factors of meeting the specified demands for fertilizer and for moisture of a given plant are both immensely important, not only in the production of crops of fine quality in abundance, but in helping to create the natural resistance of the plant to disease and even to insect attack.

During the many years of observation and experimentation at Malabar, two facets of the whole vast problem of disease and insects have emerged very clearly. One is the importance of sufficient nitrogen for *all* crops, and the other is the part undoubtedly played by the beetle and all his scavenging relatives in the general, universal pattern of birth, growth, death, decay, and rebirth, upon which the continuance of life itself upon this earth is founded.

More and more the vital role played by the element nitrogen in the growth and health of plants and finally of animals and people has become apparent, even to the point where we at Malabar have reached three additional conclusions: (1) that by far the greater part of crops raised in this country suffer from nitrogen deficiency or are actually nitrogen starved; (2) that nitrogen plays an immensely important role in the health, rapid growth, and resistance in plants to disease and even insect attack; (3) that nitrogen also plays a large role in the production of high protein content in all food crops.

Early in our work, we began to doubt certain rules once generally accepted regarding nitrogen. There was, very early, evidence that it was *not too much* nitrogen that brought about the heavy vegetative growth in small grains, resulting in poor grain production and the lodging of crops, thus making the harvest difficult or impossible. The lodging condition was not produced because there was too much nitrogen in the soils, but because there was *not enough* available potash, phosphorus, and even perhaps calcium.

Quite naturally we came across these ideas as, in the past and even to some extent at the present time, we have turned in and composted heavy sods resulting from five, six, or seven years of growth in a mixture of alfalfa, brome grass, and ladino clover. Over such a period of time and in such a long rotation the residue of nitrogen fixed deep in the soil by the leguminous plants provided a really immense store of nitrogen, and in growing wheat, barley, or oats in soils containing so heavy a nitrogen content, the lodging was very heavy and damaging.

The small grains in the field *looked* magnificent up to a certain point where the developing heads of grain turned out to be on the small side and poorly filled out, and under the first heavy rain the rank growth of stem and leaf collapsed. It was only after we began to increase greatly the amount of potash and phosphorus fertilizer on such fields, that we began to discover the real answer. With continually increasing amounts of available potash and phosphorus, the stems grew stronger and of more solid growth and displayed less and less tendency to collapse. More striking, however, was the effect of the additional potash and phosphorus upon the yields of the single heads of grain and consequently of the entire field. We began to develop heads of wheat approximately three times the length of the average wheat head in Ohio and with the heads filled out perfectly. Many heads actually measured six inches in length and many sometimes more. Winter barley heads increased in size in relation to that of the average wheat head. We have eventually reached a point where we use from five to six hundred pounds to the acre of highly available potash and phosphorus on these fields where the content of natural nitrogen is very high.

Our heavy yields from these nitrogen-rich fields do not come from heavy seedings of grain, for we use less grain per acre in small-grain seedings than most farmers, but from the immensely increased yield per head and per stalk. The same heavy yields indicate also higher protein and nutrition content, a factor confirmed by tests. (From time to time we have sold wheat to milling companies at a premium price because of the high protein content and on a basis not of *our* tests but of their own.)[7]

All the evidence over the years has pointed to the fact that our rankly growing small-grain crops lodged and gave indifferent production *not* because there was too much nitrogen, but because there was not *enough* available potash and phosphorus. In other words, one available element had gotten out of balance with the others and so produced for a given crop a lopsided nutritional balance which made impossible the optimum result in quality and quantity. This same *un*balance on the side of calcium in growing of potatoes is largely responsible, I think, for the generally accepted belief that potatoes do best on sour land with low calcium content. The case of excellent potato production on soils with a pH as high as seven to eight is described earlier in this chapter. The whole matter of proper "feeding" balances for a given plant is just as important in agriculture as a proper feeding program, varied according to production and quality needs, is in the whole parallel field of livestock.[8]

Recently, since we are able to get very high production per acre on corn, we have returned to growing this valuable but expensive grain crop, and have used corn rather than small grains as the first crop to plant on soils after the plowing up of heavy alfalfa, brome grass, ladino meadow sods. Corn, as most farmers know well, has very nearly an insatiable appetite for nitrogen, and for corn, the nitrogen in these saturated soils

[7]It is also notable that on two occasions the weight per bushel of Malabar wheat at the mill brought exclamations of astonishment and pleasure from the veteran weighers. One observed, "Good Lord! This is the first sixty-pound-to-the-bushel wheat I have weighed in more than a generation"—a statement which indicated the plumpness and filled-out quality of the individual grains.

[8]With good farmers the old practice of using one general formula of chemical fertilizer for all crops has largely been abandoned and more and more highly specialized formulas suitable to special crops has come into use. It used to be that to most farmers "fertilizer" meant simply the common 3-3-12 formula. Until very recently it was still difficult to get specialized formulas beyond a very narrow margin, and it is still difficult for the average farmer to get a wide variety of formulas, or those including minor elements, even from the enlightened Farm Bureau.

was not out of balance, although greatly out of balance for small grains. Indeed, for corn we even add quantities of nitrogen in a 10-10-10 formula to the natural nitrogen already present in the soil, with excellent results not only in overall production but in the general high-protein content of grain and stalk. As many a visitor can testify, the corn grown at Malabar in these days is so dark green in color that on a cloudy day it appears almost blue, which is how good, healthy, productive corn should took. The yellower the leaf and stalk of the corn, the poorer the yield and the nutritional quality.

Throughout the Corn Belt there has been an increasing complaint since the introduction of hybrid corn that the protein content of the corn is steadily declining. Surveys and tests have supported the complaint, but I do not believe that hybrid corn is in any way whatever responsible for the decline in protein content; its fertilizer demands are no different from those of the old open pollinated corn and are probably no heavier, save in the case where very heavy planting per acre is practiced. The decline in protein arises, I am convinced, from the general decline and even exhaustion of the available nitrogen in the heavily farmed Middle Western Corn Belt. We should never forget that proteins are simply nitrogen in another molecular form, and the two have a very direct relation to each other.

Another important factor does affect the situation—that the decline in *available* nitrogen, either in natural or in chemical form, is closely related to the fairly rapid deterioration of soil textures in the Corn Belt area and to the rapid destruction of organic material through the constant heavy fitting and cultivation of row crops without sufficient attention to the replenishing of the organic material. Nitrogen itself is closely bound in with decayed organic material, and its availability even in chemical form is largely dependent upon good soil texture and the capacity of soils to absorb and conserve rainfall. The lower in organic content these Middle Western fields become, the poorer the soil structure, the lower becomes the availability of nitrogen or any element in all its forms.

There are, we believe at Malabar, many old misconceptions regarding the use of nitrogen, and nearly all of these misconceptions are founded upon failure to observe the vital element of mineral and elemental balance. There is an old legend that nitrogen is unnecessary for the production of beans or indeed for all legumes because this family of plants produces sufficient nitrogen on its own. Many an old-timer will tell you that the use of nitrogen fertilizer on a bean crop will produce an exuberant growth but a poor yield. Such a result is conceivable, but only where the balance is poor and there is *not enough* potash and phosphorus. We have found, year after year, that nitrogen fertilizer, in the form of green manures, of barnyard manures, and even in chemical form, actually increases the yield and certainly the quality of all beans, *provided* the nitrogen is in proper balance at a *very high level* with the other necessary elements.

To put the question of this balance quite simply, it is, in our experience at Malabar, that one can have the proper balance of available minerals and elements at any level and quantity—low, medium or high—and that proper balance will tend to stabilize and increase the yield and quality of any given crop, where the crop's own individual appetites and needs are considered in the fertilizer program. However, the benefits will be increased up to optimum production in exact ratio to the amounts of fertilizer and fertility, *provided* these are constantly kept in good balance for that given crop.[9]

In our experience, there is every evidence that abundant nitrogen is in itself an efficient insecticide and that a deficiency of nitrogen tends actually to invite the attack of the scavenging beetle family, and in general of certain varieties of insects which live and thrive by sucking the juices of plants.

At Missouri State Agricultural College, it was discovered that in general many insects showed an aversion to all plants grown in good, balanced, living, productive soil. In several experiments it was found that a deficiency or poor balance of any element tended to encourage attack by these insects. The most striking example was the case of chinch bugs and corn. It was observed that the greatest infestations always occurred in the parts of a field or plots where the yellowish color of the corn indicated a

[9]It is quite possible, of course, to employ chemical fertilizers in such great amounts that they will actually become damaging to the crop, especially in hot dry weather and in soils of low organic content and poor texture. It is also possible, for example, to use superphosphate in such great quantities and for so long a period of time that the sulphur content of the soils will become toxic. One assumes, however, optimistically, that the elements of common sense and potential observation exist in most men and that most gardeners and farmers have had some experience with soils. I have also heard the assertion that the use of minor elements in fertilizers may in time produce a toxic condition in soils. Such a development is entirely possible but, assuming that the amounts used are sensible and practical and sufficiently small, this toxicity does not appear to create any serious risk. Also, most plants exercise a certain selectivity, utilizing an amount of any available given element or mineral up to its needs and not beyond.

The loss of these minor elements through leaching into lower levels of the soil is also considerable and should be calculated in any real, sound, and continuing fertilizer program.

deficiency of nitrogen; on the deep green parts of the field where there was abundant nitrogen, the insects actually showed a distaste for the corn.

In Kansas similar irrefutable evidence was turned up in the case of the green bug and wheat, and it was discovered that, as a protection against attack, a top dressing of nitrogen on a yellowish nitrogen-deficient wheat field (provided there was sufficient moisture) was as effective or more so than heavy dusts of inorganic poisons. It is notable in this case that the worst attacks of the green bug, one of the aphids, occur during dry or drought years when insufficient moisture tends to make whatever nitrogen is present in the soil largely unavailable.

In the cases both of the chinch bug and corn and of the green bug and wheat, the clue to the value of nitrogen could have been observed long ago if the research had been sufficiently thorough, and doubtless it was observed long ago by good, intelligent farmers here and there. The concentrated attack of the chinch bug upon limited, nitrogen-deficient areas of corn has been going on for generations, and any passerby observing one of the vast fields of the wheat area could easily discover that the green bug does not attack a field by entering at one side and moving across the field like an army, or indeed like an infestation of army worms, but settles on the yellowish nitrogen-deficient spots and works outward from them in a consistent circular pattern.[10]

In our own plots at Malabar, where the level of natural organic nitrogen is far above that of the average farm or garden, we have also found that even more nitrogen could actually serve in the role of an insecticide. This proved especially true in the case of melons, cantaloupe, cucumbers,

[10]On one occasion on an enormous Texas ranch, the author observed a striking example of the relationship of green bug attack to nitrogen. In a vast field of eighteen thousand acres the author drove with the ranch manager over a newly developed road. By the side of this road a strip of deep green, healthy wheat, perhaps ten to twelve feet in width, ran the whole length of the field. The marked difference in vigor and color of this narrow strip from the rest of the field raised the query as to whether a test was being made by the use of extra amounts of fertilizer. The answer given was, "No." The rich green wheat was growing on what, since the very opening of the ranch, had been the old road, and therefore was still virgin soil. The ranch had simply constructed the new road alongside the old one and plowed up the old road. That rich green strip was striking evidence of what the fertility of the whole field had once been, and the pale, thin wheat growing elsewhere was equally striking evidence of the degree to which that fertility or its availability had declined. The green bug had already begun the attack in circular patches throughout the field, but there was, so far as I could discover, not a single insect on the narrow, rich green strip that paralleled the road. In a way this experience supported my general belief that much of the increasing damage done to American crops by disease and insects is caused by poor agricultural methods in the past and the general slow depletion of soils and a steady decline in the availability of fertility.

and squash, where an additional side dressing of chemical nitrogen, promptly watered in, served to drive off the attack of beetles, which did not return. This general family of plants, however, is one that is especially susceptible to bacterial wilt, which is one of the most difficult of diseases to control and which, apparently, is distributed by the beetle during its first attack on the seedling. For this reason, a rotenone dust is sometimes used almost immediately to check *any* beetle attack and consequently a spreading of the disease.

At this point it might be observed that at Malabar we are inclined to be reluctant and to hold back in the use of *any* dust or spray until absolutely necessary, even in the case of such mild vegetable poisons as rotenone, nicotine, and pyrethrum. These may indeed be virtually harmless to humans and to animals, but they are known to have a strong toxic effect upon insects and in the case of rotenone even upon fish.[11]

It is therefore reasonable to suppose that even a mild insecticide such as rotenone could also have a serious toxic effect upon certain living soil bacteria and even upon earthworms and other beneficial living organisms which are essentially a part of any truly healthy, living, and productive soils. If this could be true of a comparatively weak vegetable poison, how much more destructive to good soil conditions must be the effect of arsenic and high-powered inorganic dusts and sprays which are used wholesale today. It is not impossible that the resistance factor in many commercial, vegetable-growing operations has been greatly reduced through the destruction by wholesale use of poisons of the living organisms of the soil which are a part of any really sound, permanent, and healthy agriculture. Such a condition could easily set up a vicious circle in which the steady destruction of living soil organisms gradually diminished the availability of fertility and served in turn to reduce the capacity of the soil to produce fast-growing, vigorous, resistant plants. In order to correct this weakened condition, more and more poisons in increasing amounts are then used, and in turn accelerate the whole process of reducing a good soil to a condition of sterility in which it becomes merely a sterile *medium* in which to grow things, instead of a living and healthy soil in which the process of birth, growth, death, decay, and rebirth is in constant operation. Some such vicious circle may well account for the

[11]The discovery and employment of rotenone as an insecticide arose from its use among South American Indians in the capture of fish. A small amount of the poison in a pool or stream served to stun the fish and make them easy to take. Rotenone is used also in certain state-managed programs for clearing ponds of coarse and valueless fish before restocking.

increasing infestation of disease and insects, and declines in production and quality on many hard-worked vegetable production and farm areas throughout the country.

SUMMARY: In all our experimentation and research at Malabar in the checking or correction of attack on plants by disease and insects, we have consistently followed a reasonable course, making no claims which could not be verified at any time on the plots and the vegetable gardens and the farm itself, all of which are open at all times to all the public. We have never made any unqualified, intransigent claim that all insects and disease can be controlled wholly through the creating of sound, living soils, or that both disease and insects and their attack can be entirely eliminated, although in some cases, as in that of celery and blight, or the attack of potato bugs, this has seemed to be proven. At least we are no longer troubled by either, and certainly this was not because the crops were grown on perfectly clean and uninfested soils; exactly the opposite was true.

We have striven constantly, however, to attach the factor of resistance to the proper kind of soil and upon the healthy, rapid, vigorous growth of the plant. In most cases we have succeeded in creating a resistance which certainly prevents the attack, either by disease or insect, from becoming a catastrophe, which would mean total or near-total failure of a crop. In most cases the incidence and violence of the attack has been reduced year after year to a point where in many cases dusting or spraying has become unnecessary or where, as in the case of the bean family described in this chapter, the attack has come upon the *dying* plant *after* the crop has been harvested and is therefore of no importance either to production, quality, and nutrition or to economic factors save that the labor and materials involved in any process of dusting or spraying are eliminated.

Certain convictions and some theories, as yet unproven, have arisen from our close observation of conditions and from the carefully planned experimentation. One of them concerns the place played in the whole picture of agriculture and horticulture by the insects of the voracious beetle family.

This family belongs fundamentally to the whole range of factors which are concerned with the breaking down and the reduction of all dead materials into soil and consequently to the constantly renewed basis of fertility, life, and reproduction. Among these countless elements are bacteria of all sorts, fungi and molds, earthworms, and other living soil organisms; without these elements the surface of the earth would by now be cluttered to a depth of many feet with dead material which had not been broken down and turned into soil, fertility, and the potential of reproduction and continued life. With this fact in mind, it seems reasonable to suppose that the scavenging beetle family of insects, in carrying out its mission in the pattern of the universe, has an actual distaste for whatever is healthy, alive, and growing vigorously, and has a definite *preference* for the sick, the dying, and the dead plant. At Malabar and indeed even in a backyard garden there is always ample evidence that such a theory is valid.

At Malabar we do not claim that we can protect through the soil, plants which are attacked by the larval form of certain butterflies and moths such as those larvae which attack cabbage, tomato, and tobacco plants. These larvae are born of eggs deposited by butterflies and moths which themselves never attack or eat the foliage of these plants. The larvae are left willy-nilly where the eggs are deposited to feed upon what is at hand and most suited to their tastes. It is apparent that the parent moth or butterfly is attracted by some means, perhaps a sense of smell, to deposit her eggs on or near the plants which are particularly suited to the tastes and needs of the larvae. However, we have found that these do not exist in vast numbers in the Malabar plots and that a single light dust of rotenone is likely to prove completely effective in exterminating them. It is quite possible that in this field of larval infestation, the birds have played a far greater role than we have been able to measure, and so reduced, locally at least, the general population of the egg-laying moths and butterflies.

Nor do we pretend at Malabar that we can protect a crop through the soil from attacks by such voracious enemies as the locust, the grasshopper, and the armyworm, which at times can overrun crops by the million and the billion. In the case of such invasions by insects of the category of the locust, which, failing all other sources of food, will consume fence posts, it is obvious that the most drastic measures must be undertaken, measures unjustified under other conditions.[12]

[12]In 1954, partly in jest and partly as a challenge to certain tobacco-growing friends in Kentucky and Virginia, we predicted that we would produce a record yield of tobacco both in quantity and quality without dusts and sprays, without weeding, and without cultivation. We raised two varieties of Burley and one of Connecticut tobacco and came within a hair of carrying out our boast completely. Photographs of the tobacco are contained in this book and speak for themselves. The tobacco plants were set in mulch, were

Largely speaking, in the vegetable gardens, the plots, and the fields at Malabar we have sought and are still seeking ways of producing optimum yields in crops; optimum in quantity, in quality, and in nutritional value, with the lowest possible expense in labor and in dusts and sprays. There is every evidence that in accomplishing such a goal we can at the same time reduce the attack of disease and of some insects far below the degree of incidence commonly encountered on farms and in market gardens throughout the U.S. under general existing practices of agriculture and horticulture. We have evidence that in the case of *some* vegetables and of *some* plant diseases, virtually total immunity to attack can be achieved without the use of any dust or spray whatever.

Secondarily, in the case of vegetables and of all foods including milk, we have no liking at Malabar for consuming in our daily meals quantities, either large or minute, of poisons universally recognized as lethal, or of poisons such as arsenic or DDT which the system does not eliminate in any normal fashion, but which accumulate gradually and slowly within the human body. Nor do we have any desire to act as laboratory specimens for the testing of viciously poisonous inorganic chemical, byproducts dumped on the market without proper tests or research into their lethal quantities, poisons advertised as so violent and viciously effective in destroying insect life that the operator is warned to use a gas mask while handling them. Whether they do real harm and serve in a general way to impair the health of the whole nation and to create an increase in the degenerative diseases of middle age, I do not know at this stage of the game, nor does anyone else. One thing is certain—that used as they are in the production and processing of our foods to the amount of millions of pounds a year, they can do no one any good.[13]

---

never weeded or cultivated, were given soluble fertilizer through the irrigation, and never received a dust or spray. The Connecticut tobacco was attacked by about ten or a dozen tobacco worms which were picked off, although no worm appeared on the Burley, despite the fact that it was growing in the next row to the Connecticut tobacco with the leaves brushing each other. Why the worms practiced this selectivity I have no idea.

[13] The accumulations of poisons in soil where they have been used year after year as dusts and sprays can be staggeringly large, even to the point of saturation, at which the growth of the trees and plants themselves are damaged and impaired. In Washington State, before laws were passed forbidding the use of arsenic compounds for dusting or spraying fruits, accumulations of as much as two tons per acre of arsenic compounds were found in the soils of some orchards. The wholesale use of the various weed killers of the 2-4-D variety may in time tend to reduce or even perhaps prevent the production of certain crops. In the early days of Malabar we employed 2-4-D to clean an entire farm of bindweed and wild artichoke. Half the farm was given two sprays of 2-4-D in water emulsion, the other half was given the same amount of spray in oil emulsion form. Two years later we found

it virtually impossible to get an alfalfa seeding on the portion sprayed with the oil emulsion, and even today, five years afterward, there is still evidence that something in that
particular area is damaging to young alfalfa seedlings. The line between the two is strongly marked, with a good seeding on the side where the water emulsion was used and a poor
one on the side where oil emulsion was employed. The only deduction we could make was
that the 2-4-D in the form of water emulsion had quickly leached out of the soil or
become rapidly impotent, while that used in oil emulsion clung to the soil particles and
the organic material and remained in the soil, and remained active, for a long time after
being used. In New Jersey I have seen established orchards in which it was no longer possible to grow grass beneath the trees because of the high arsenic content of the soil itself.

## THE HARD-WORKING SPRING
## AND THE HOUSE NOBODY LOVED

THE SPRING HAD BEEN THERE FOREVER, PROBABLY SINCE THE TIME WHEN the last melting ice of the second great glacier receded from our country. It comes from the deep Silurian sandstone, one of the oldest formations in the world, flowing up through one of the many crevices created ten million years ago when the huge weight of the glacier cracked the heavy sandstone underneath. It comes out of the rock inside a natural cave, and from there flows steeply down the hillside to the old springhouse with huge troughs made more than a century ago by hollowing out great blocks of sandstone taken from the nearby outcrop of rock from which the spring itself emerges.

The big old Bailey house was built on its present site because of the great spring. The house itself sits just below the outcrop of multicolored sandstone and was built from bricks burned on the spot. It is a handsome house, resembling the big brick manor houses of Maryland and Virginia, truly an astonishing house in its style and grandeur to have been built on the frontier in the midst of the beautiful Ohio forest not many years after the Indians were defeated at the Battle of Fallen Timbers and Ohio was opened up to settlers.

The house stands not far from what was once the important settlement of Newville, the largest community of that early frontier world, and the center of life in the county. Through it, running northward up the headwater valleys of the Muskingum River from the Ohio River to Lake Erie, there ran a road which had once been an Indian trail. As the country opened up after the final conquest of the Indians, it became a road, not a very good road and little more than a means by which the wagons carrying wheat, the sturdy coaches, and the driven cattle and hogs could make their way. And so as more and more wagons and travelers and cattle moved on over the road there grew up a necessity for taverns, or merely places for the drivers and coach passengers to spend the night. Here and there a settler raised a house big enough to take in passersby for the night. They were at the same time rather imposing residences and had a couple of large rooms for the accommodation of travelers. All the men slept in one room and all the women in another. Frequently a traveler spent the night in a double bed with a companion he had never seen before and possibly would never see again. In Kentucky there is many an old farmhouse with an outside stairway running to a room on the second floor which has no access to the rest of the house. Such rooms were put to a similar use; a traveler might eat at the family table and retire for the night to his room by means of the outside stairway. It was a good arrangement in half-wild country where one out of three passing travelers might well be a suspicious character.

For such a house and tavern as the one at Malabar, the famous spring offered an ample supply of clear cold water for all purposes and so the original settler, David Schrack, erected the imposing and beautiful house and tavern which still stands just below the spring on the edge of the road.

Since the road first came into existence it has gone through many vicissitudes. First it was a rough but important highway passing through the busy main street of the town of Newville, and from thence upward and northward to Sandusky on the lake. Then another settlement called Mansfield, nearer the center of the county, was made the county seat and an imposing courthouse erected, and the once flourishing town of Newville went into a decline. The coming of the railroads, which passed through the new county seat of Mansfield, settled the fate of Newville, and the older settlement went into a decline from which it never recovered. The road, which had once seen the busy passage of great wagons, coaches, and herds of cattle and sheep and hogs, ceased any longer to be important and presently became merely a muddy or dusty backcountry road through the beautiful half-lost valley. It acquired the name of

Pleasant Valley Road, which still designates it on the new and shining enameled signs the county commissioners have put up to guide travelers among the countless wandering, unnumbered roads of our half-forested hilly countryside.

The death of Newville finally came in the thirties when the now famous Muskingum Conservancy and Flood Control District came into existence and a dam was built to impound flood waters which might have engulfed the village itself. The death came easily, for the community by that time had become merely a dying relic of the old frontier, with ruined and empty houses and a half-ruined "hotel" that had gone unpainted for years. As a small boy I remember the village well when it was already a kind of ghost town with a singular faded nostalgic beauty. When the end came, the few remaining citizens moved their houses up the hillside and the once flourishing town became merely a series of gaping stone-walled cellars or was obliterated altogether by the plows of neighboring farmers. It was on rich bottom land that was easy to farm. Meanwhile, it seemed that the Pleasant Valley Road had gone back almost to the insignificance of its days as an Indian trail. It seemed to lead nowhere at all but to the beautiful hill farms which, in the days of horse and buggy, were almost lost in the surrounding forests.

Then with the establishment of the Muskingum Conservancy District and the creation of a big lake and a wonderful forest park recreation area, the road came to life again. Tourists, travelers, fishermen, hunters, mothers bringing their children to the beach, sailing and motorboat sportsmen began to flood the old Indian trail once more, in a constant stream of traffic during the summer months. And presently many an industrial and white-collar worker who had built a cottage or cabin in the area for weekends and holidays found that it was very pleasant to have gardens and even small farms of their own, that it was good for the children to live on the edge of the lake and the forest with swimming and fishing and boating and hunting in one's backyard, and in greater and greater numbers they began to desert the neighboring towns and cities and to transform their weekend cabins and cottages into permanent homes, living the year round in the beautiful area and going back and forth to work in the cities.

And so Pleasant Valley Road became once more a busy highway, no longer traveled by Indians, drovers, wagoners, and coaches, but by automobiles; every other one towing a boat or a trailer. After them came the tourists and the people who came from every part of the world to study the beautiful Muskingum watershed development, and frequently to pay us visits at Malabar.

The old residence and tavern was solidly built and it still stands beside the road that has become noisy and busy once more. Its walls are nearly two feet thick, built of bricks which were burned from the shale to be found in the neighboring fields, and the whole big house rests on a rock-solid foundation of huge sandstone blocks carved out of the low cliffs from which flows the big spring. Some of these great blocks of stone weigh in the tons, and in the high ceilinged cellar there is a vast fireplace, once used perhaps at butchering time, so imposing that I can stand in the opening. The cellar itself extends under the whole of the big house and as the house is built on a hillside, one can enter the cellar on the lower side directly from the road.

The original builder, I long ago divined, must have been a man of determination and character, who built solidly for the ages, and when I went to the quaint *County History*, published in 1880 when there were still people alive who remembered Newville as an important town and farms that were still being cleared from the thick, rich forest, here in part is what I discovered concerning the builder of the big brick house.

SCHRACK, DAVID (deceased) was born in Center County, Pennsylvania; was of Scotch-English descent and a farmer by occupation. He purchased of Thomas Pope a quarter-section of land in the southeast portion of Monroe Township and moved thereon with his family. He subsequently entered the quarter-section adjoining his first purchase to the north; on the Pope farm there were about four acres cleared and a rude cabin built thereon. Mr. Schrack and family lived in this cabin till they were able to put up a more comfortable and commodious house. His farm was covered by a dense and heavy growth of timber, and required a vast amount of hard labor to prepare the lands for cultivation. And though the soil was rich and productive, his grounds were stony and hilly; but by hard labor and perseverance on the part of himself and his sons these difficulties were all overcome. Mr. Schrack lived to see a massive brick dwelling occupying the place of the rude cabin. He lived to see the dense forest give place to fruitful fields. He lived to rear a large family of children, and to become comparatively wealthy. Mr. Schrack was a "mighty hunter" and many were the deer, wild turkeys, and other wild animals that fell beneath his unerring aim. He also shot quite a number of bears when he first came into the country. He was among the Indians a great deal, with whom he was always on friendly terms. Mr. and Mrs. Schrack were the parents of fourteen children, three sons and eleven daughters.

On reading all that one is tempted to lean back and say, "What a man!" Moreover, his great-grandson, Charlie Schrack, who is our next-door neighbor and one of the best farmers of the county, told me that his forefather had often taken a deer he had shot, put it on his back, and carried it to the town of Wooster, forty miles distant, and returned carrying the hundred-pound bag of salt he received in return for the deer ... and all the way over up and down hill over rough deer and Indian trails.

Of the original owner, Thomas Pope, there is no mention in the old history book, and apparently there is no information today available concerning him, but it is likely that he was an officer in the War of the Revolution and the War of 1812, for the farm and the big old house lie in what were the Military Lands, in which land was given to old soldiers by Congress as a reward for their services in behalf of the nation. It is just possible that Thomas Pope, whatever his history, never even saw the forests, marshes, and hills of Pleasant Valley. But the indication is that David Schrack, when he came into Ohio, was a person of substance, for he *bought* his land from a former owner and did not receive it from the government merely for the taking as did many early settlers.

In the brief biography there is so much contained in so little; when it refers to "a dense and heavy growth of timber" only an Ohioan can understand how much is contained in the simple phrase and only an Ohioan from our part of the state, which is still mostly forest country, can realize how dense and heavy that growth can be ... a growth, in David Schrack's time, of vast primeval oaks and chestnuts, beech and maple tangled with wild grape vines, and an undergrowth of saplings and ferns and berries that is impenetrable and is like a Brazilian jungle in the season of rain. Even today, with heavy machinery and bulldozers, one hesitates to tackle the problems of clearing away such forest or even the pallid second- and third-growth forest that covers our roughest land. Yet David Schrack and his sons cleared away that massive forest with their own hands and the aid perhaps of a team of oxen. I never walk over those fields cleared by David Schrack and his sons without a feeling of awe and reverence for the man who has been dead for over a hundred years, or without reflecting upon the pride he must have felt on the day he moved his family from the "rude cabin" into the handsome manor house-tavern which now belongs to us, and which all of us at Malabar love so much and treat with actual reverence and devotion.

In the second generation, the farm left the hands of the Schrack family; they married or went away or purchased land that was easier to farm, and the place, which we now call the Bailey Place, fell into the hands of a series of owners who neither loved nor cherished it, and a series of year-by-year, nomadic tenants who ruined the farm and wrecked the inside of the old house that was the pride of David Schrack. Sometimes they were ignorant, sometimes lazy (one at least is on record as a half-wit). They were representative of a period when people said, "Anybody can be a farmer," and men who failed at all else turned to the land for a meager living.

Through the house for years there wandered a procession of down-at-the-heel people, always unfortunate, and many of them unfortunate through their own defects. Not one of them, clearly, ever possessed the hardiness, the character, and the determination and pride of David Schrack. Few of them had even human dignity.

They tore out the old and beautiful winding stairway that led upward from the fine hall and burned the wood, along with that of the mantel-pieces, as firewood. They failed to repair the roof and moved from one room to another as the water came through. They chucked the cans from the tinned goods they bought at the grocery out the windows because they were too shiftless to raise a garden. The beautiful house, which is a kind of monument to the hardiness and dignity of David Schrack, became very nearly a shabby, abandoned ruin. The final owner before us made a valiant effort to farm properly and clean up the place and repair the house, but by this time the ruin was so great that he found the task beyond his powers, sold the land and house to us at Malabar, and went elsewhere onto flat land that was easier to manage.

But the big spring is still the thing which gives value and beauty and charm to the whole of the farm. For centuries, far back into the shadows of the remote past, it was known to the Indians. It was frequented by wild animals, and especially by the deer which still come morning and evening to drink at the pond across the road, because even in the pond the water is cold and clear. In summer when we cut the hay we find their trails through the tall grass and alfalfa leading from the Jungle, a big area of swamp and forest which we have left in primitive condition for the wildlife. And the pond is teeming with fish, and in spring and autumn its transparent surface is covered with migrating wildfowl of every kind. The water, which comes from deep in the ancient sandstone, is soft as rainwater and the women who wash the vegetables in the big stone troughs like working in it because it softens and beautifies their hands like any beauty cream.

The old farm lies in the very heart of what was Johnny Appleseed's country, and he visited the spring countless times. It is pleasant to think that Johnny and David Schrack must have known each other well and frequently met at the spring and that Johnny must have slept many times in the plain old barn with its huge wood-pegged timbers of black walnut and white oak. Johnny always refused the comfort and shelter and good bed of a house. He preferred to take himself, with the cooking pot he wore as a hat, with his "poke" of fennel and apple seeds, to sleep in the barn in the great mow above the beasts.

Occasionally there rises in my imagination a pleasant picture of David Schrack and Johnny Appleseed drinking and talking at the big spring, together with their Indian friends, and sometimes a fourth figure enters the picture—that of the Lost Dauphin who, legend has it, once wandered through our country with Johnny Appleseed … a plump young man, raised among the Indians, dull-witted, like so many Bourbons, and simple, with the marks of scrofula on his throat and the confused vague and misty memories of great mirrors and torches and soldiers and mobs of screaming men and women. There are still old people in our countryside and even in France who believe fervently that the Lost Dauphin was in truth the boy spirited away from the Temple and the true King of France.

Old houses have an aura of their own, as if the spirits of the past had somehow left behind them some of the essence of their very lives and characters. When we first took the Bailey Place and the big old house in all its ruin, the whole of the air about the place seemed infected with weariness, discouragement, and despair, but slowly, as the place has changed and love been spent upon it, the aura has changed, as if we had somehow driven off the spirits of all those who for generations abused both the land and the house and had no love for either. It is somehow as if we had summoned back David Schrack and all his friends, and driven off the spirits of all the long procession of shiftless, luckless families who passed one after another through the place. I think today that it is the spot which all of us at Malabar love best. The spring, the cold running water, draw us back again and again. It has a kind of beauty quite different from the high, wild, solitary beauty of the Ferguson Place. The spirits that now gather here by the spring and David Schrack's "commodious mansion" are of a special sort who would have been friends. I think they loved eating and drinking, making love, working, hunting, and fishing and all the good things with which the Good Lord provided us in a world that was meant to be enjoyed.

The spring had many vicissitudes and has seen the Mound Builders who preceded the nomadic Indians, and after the nomads the agricultural tribes which built cabins and raised squash, corn, and beans in the valleys of our country, and presently the *coureurs de bois*, those French and half-caste trappers who never settled anywhere, but made a living by hunting and trapping the wild animals for the fur trade. It is sometimes pleasant to imagine that the mink, the muskrat, and the ermine which adorned the ladies of the courts of Louis XIV and Louis XV and the furs worn by Jean Jacques Rousseau and French Ambassador Benjamin Franklin in their portraits may have come from the marshes and forests of our own Pleasant Valley. And finally the first English-speaking settlers like David Schrack. Always, as is the case of springs and fountains in all countries everywhere, it was a gathering place, alike for people, for cattle, and for wild animals.

The speculation as to what our countryside would have been if the whole of the great Mississippi Valley had remained French and been settled by the French is irresistible; certainly it would have been a different country and even a different civilization, quite possibly a better one in many ways. But the French have always been poor colonists and never arrived anywhere in sufficient numbers. They had and still have the most beautiful of countries and the most enlightened of civilizations. One can understand why they are unwilling to leave France for the hardships of the frontier and the barbarities of other "cultures."

As a boy, when I used to ride the county with my father in a buggy drawn by a team, I remember the great trough, hollowed from an enormous oak tree, that stood on the edge of the road beneath a huge and ancient willow tree—a tree wracked by the scars of ancient storms, from which sprang eternally the fringes of delicate young shoots of new growth; a tree which had died and been reborn a dozen times. Here travelers stopped to water their horses and refresh themselves with the clear cold water. All around the leaking old trough grew mint and black peppermint, ferns, and wild yellow iris, and in the cold clear water below it the peppery watercress which my father and I always gathered to bring home with us. It is probable that David Schrack himself placed the great oaken trough there by the side of the Pleasant Valley Road when Newville was a thriving frontier market town and Johnny Appleseed had his nursery of fruit trees on the edge of the marshes which are now covered by the waters of Pleasant Hill Lake. I like to think that the old trough which I knew as a boy was a kind of bond between the two of us, David Schrack and myself, who both have loved the farm and the spring.

In the generation in which I was away from the county and the valley, the old oak trough rotted away and the ancient, apparently immortal willow was finally slain forever by the ax to make room for the widening of the old stagecoach road. When I returned it was gone and in its place was a circular fountain of cast iron, picked up perhaps from a junkyard where it had found its way from the lawn of some house of the U.S. Grant period ... a lawn where it had kept company in its day with cast-iron deer and dogs and circular or crescent-shaped beds of kolias, salvias, and geraniums. It did not belong there on the roadside close by the purity and dignity of the old tavern built in the simple style of the Greek revival, and one of our first acts after we acquired the farm was to remove it and its implications of ugliness, ostentation, and vulgarity.

The fountain was succeeded by a small pond surrounded by wild flags and marsh plants, but the pond was never a very great success. I suppose that I was, without understanding the motive, trying to re-create the ragged wild beauty of the rotting old wooden trough and the weather-scarred ancient willow. But I never managed to achieve that beauty of cold spring water dripping over the moss and rotting wood into the beds of watercress below. In summer the little pond was likely to be filled with algae and heavy-growing water mosses which choked it. The water was too cold for pond lilies, although the cress flourished at the spot where the clean spring water entered it. In winter the wandering muskrats took possession of it and burrowed in the banks and in the dam which sustained it. The water leaked across the highway and froze, creating a menace to speeding cars passing on the highway. Definitely it was a failure ... even an eyesore.

Yet I was tormented by a desire to create about that stream of clear cold water something which was at once beautiful and useful as the old wooden trough had been long ago when it provided water for the sweating horses and tired hot cattle which passed that way on a hot summer day. I contemplated restoring the old trough, this time in stone, but such a plan seemed an affectation and an absurdity in a day when horses and cattle no longer passed along the old stagecoach and cattle road. In the end, in the simplest and most natural way, the solution was found, which was at the same time simple, beautiful, practical, and even functional. It grew naturally out of the character, quality, and needs of Malabar Farm and of the times in which we live, exactly as the original trough had been set up because it was needed and so was used.

One knows when a house is loved and when it is not loved. One can tell by the eaves, by the stairways, by the shrubbery, by the very grass which grows around it. This house beside the great spring had not been loved in generations. Perhaps it had not been loved since David Schrack, the Great Hunter, died at last and left it together with the spring and the hilly farm to children who sold it and went elsewhere to farm land that was less beautiful, but was flat and easy to manage and grew crops easily without any special love, intelligence, or knowledge. For years it had gone from owner to owner, some of them speculators, some of them men who could buy it because they had very little money but not enough to purchase a better, more prosperous place. As if the farm and the spring and the old house resented them, they were each one in turn defeated, along with the dreary line of yearly tenants whose only thought and only motive was to steal from the house and the farm all they could during the brief period of their tenantry. And all the time, I think, the house and the spring wanted to be loved before it was too late and the house fell into ruin so terrible that it had to be destroyed.

When we took over the place, such a fate was not far away. A few more years and the beautiful house would have fallen apart, and because man could not destroy the beautiful spring, another house would be built there on the same site, a house that belonged to a different world, a ranch-type house perhaps with a picture window, for the view from the site encompasses all the valley in both directions, from the hillside forests to the blue lake; but it would be new and strange and without the power of old houses to evoke the past and summon up the spirits.

In the years of its misfortune, some owner had seen fit to desecrate the old pink bricks burned on the place, with a coat of dirty yellow paint. If one had sought to insult and defile its beauty, it would have been impossible to have contrived a better way. Certainly no one with any other purpose in mind could have chosen so vile and repulsive a color. It was not even the decent yellow of the good bare clay from which the bricks had been made, or the honest rich brown of good barnyard manure. One can describe the many things it was *not*. It is difficult to find words to describe what it actually *was*.

The roof of the high porch had long since rotted away, and the porch itself was so badly decayed that it could not be restored. Just across the road, blocking the magnificent view, stood a ramshackle building, unadorned by paint save for a huge faded advertisement for Bull Durham tobacco, born of the poverty of some past owner who was paid a dollar or two for the advertising privileges in the days before desecration of the landscape became the dedicated purpose of penny-snatching advertising agencies. Behind the building there was a junkyard of old and rusting abandoned farm machinery, through which the tall weeds grew in the summer. Then came the barn, also built across the view, and a sickly, unnatural marsh created by the eroded silt from the hills above, which had long since choked the channel that carried the waters of the spring downward toward Switzer's Run and the lake. The silt was so deep that the channel was actually higher than the surrounding land, and the water poured out of it into the surrounding weed-grown area, which had apparently been used for years as a dumping ground and trash heap.

As I have written before, all of us at Malabar had long coveted the whole of the Bailey Place, with the beautiful house, the wonderful spring, the view from the house, and the magnificent view of all the valley, the lake, and the forests from the top of Mount Jeez, where the abandoned fields were covered with poverty grass and broom sedge and wild carrot. But the owner had always made difficulties and, knowing that we coveted the place, had asked fantastic prices and even tried to force a sale by threatening to sell it off in small parcels to cottage builders.

Actually, he did sell two sites which came into the possession of two old gentlemen who built cottages overlooking the lake for the summer and weekends and are still happily with us. They are, in a way, Dickens characters, and one of them has developed his one-acre place to a point where there is no more room for development. He has built a pretty cottage and has a trout pond fed by one of the offshoots of our big spring, a duck pond, a barbecue pit, a fruit orchard, and hundreds of feet of picket fence. It has been so heavily planted with evergreens and flowering shrubs that it is scarcely possible any longer to see in or out of the cottage. It strongly resembles the castle in *Great Expectations*. But one thing is certain; the place is *loved*. The mere sight of it must bring pleasure to the passerby. It is bursting with abundance and fertility and life. The picket fence and the gates are always beautifully painted and white. Although the two places are like bites out of the very middle of the Bailey Place, I do not covet them because they are loved and they give pleasure not only to their owners but to all of us at Malabar and to the passersby on the highway.

I sometimes think that perhaps from another world, old David Schrack, the Mighty Hunter, had something to do with the Bailey Place coming at last into our possession, for one Sunday morning the owner and his wife, in their best clothes, called upon me and asked if I would do a favor for a neighbor. I thought at first that perhaps they wanted to borrow money, but I was wrong. When I asked what the favor was, I found that they wanted to sell me the place and that this time they *really* wanted to sell it. It seemed that they had bought another farm in the flat country on the strength of having sold the Bailey Place, without even offering me a chance at it, to a stranger to the valley. Then, after they had made all the arrangements to take over the new place, they discovered that the potential purchaser could not raise the money. They wanted desperately to get away from the Bailey Place, for it had defeated them as it had defeated everyone since David Schrack had wrested it from the wilderness. They wanted to escape. In short, they were stuck.

When I asked how much they wanted, the price was a third what they had been asking ,and in a moment the Bailey Place, wonderful Mount Jeez, the old house, and the beautiful spring all belonged to us. Somewhere in that other world, I think David Schrack was chuckling.

It has taken a long time to bring back fertility and to restore and develop the beauty of the place, and the task is by no means finished, as probably it never will be. But that is one of the pleasures of being a landowner, especially in a beautiful countryside; there is always more to be done, there is always just ahead another horizon, and in our own country, over each difficult hilltop, there is a new and beautiful small world.

The first task was to clear away the accumulated rubbish of years, to pull down the shack with the Bull Durham sign. Then came the repairs to the porch and the eaves and the roof and the windows with their old pale mauve rolled-glass panes. Then the whole house was given a coat of whitewash to cover the dung-colored paint—a temporary measure, for one day it will all be sandblasted off down to the pink color of the old bricks. The marshy wasteland and trash heap below the barn were drained and cleared up; the silted channel was deepened and cleaned and the spring water permitted once more to go singing on its way to Switzer's Run. The weeds and poverty grass of Mount Jeez was turned into good pasture for the cattle, who wander over it in the evening twilight of a hot summer day and are visible a mile away across the valley at the Big House. The sickly artificial marsh, long since drained, has become in time one of the finest and most beautiful of vegetable gardens, and each year David Rimmer and Patrick Nutt make it richer and more beautiful. In it roses and delphiniums, tomatoes and asparagus and celery grow side by side in orderly rows, as in a French *potager*. And the house itself has been partly restored, although the full job of putting back the beautiful stairway and mantelpieces burned for firewood in the now bricked-up fireplaces still remains one of the great pleasures which lie ahead.

It is unnecessary to restore the beautiful proportions of the rooms and the fine scale of the old windows and doorways; they are still there, indestructible even at the hands of the fly-by-night tenants, who in their poverty and ignorance came to hate the place. Lilacs and flowering shrubs have been planted all about it and in the spring the slope behind the house leading up to the cliffs of sandstone is filled with clumps of narcissus and jonquils. Along the road we have put up with our own hands a high retaining wall built of the huge cut stone coming from the foundations of the old mills and houses which have long since disappeared from the neighborhood (some of them from the old cellars of the demolished settlement of Newville), and at the top grow irises, poppies, and yellow day lilies. None of it cost much money, barely more than I would spend in a day or two in the city. It was not money that was needed so much as love, a sense of values, and some of the art which the original builder, David Schrack, brought to the big but simple and classic old house.

The shrubbery (it was in reality merely underbrush) was cleared away, the rubbish carted off, and presently the house could be seen again from the road, a loved house, rich with an aura of the joys and miseries of generations, it almost seemed to straighten up and stand erect with a dignity which replaced the old sense of hurt, neglect, and shame. Sitting close to the old valley road but high above it ... so high that even on the porch one is unaware that the road is there at all ... it seems to have a new pride that is almost arrogance.

And as if my own feeling for the house had communicated itself to all the others at Malabar, friends began to contribute touches and ideas and bits of work. Kenneth and Jim Cook put up the high, long terrace wall, made of old hand-hewn blocks of multicolored sandstone, which divides the lawn from the road and gives the house an air of sitting high up but firmly upon a pedestal. They did it while I was away in California, as a kind of gesture, a kind of gift to me, but more, I think, to the old house itself. Someone, I do not even know who it was, planted an old-fashioned snowball and a common Dorothy Perkins rose just at the side of the door to the ancient springhouse. They have flourished. They were placed in exactly the proper spot, as if they had been left there by one of the passing winds which sometimes sweep down the long corridor of the valley. The bracteatum poppies, the day lilies, and the irises grow altogether in a careless luxuriance at the top of the hot dry wall as if they had chosen that very spot, and the chinks in the wall itself are filled with aubrietia, yellow sedum, hen-and-chickens, California poppies, and portulaca, all contributed by friends and even passersby and casual visitors.

And there is now a view from the windows and the high porch across the round mirror-shaped pond, the rich ordered garden of vegetables and flowers, and the fields of exuberant alfalfa and the corn which is almost blue in color from the fertility of the soil. And in the far distance across the valley lies the misty line of the forest which changes constantly with the seasons, from the delicate pastel colors of early spring when the pink budding leaves of the red oaks make the whole tree seem covered with pink blossoms like the piñeiro of Brazil, and the white dogwoods and pink wild crab apple blossoms appear along the edges like the foam of a green tide breaking on the beach, to the autumn when the whole forest seems to burst into flame. Only a little while ago, there was no pond bordered by willows, wild flags, and marsh marigolds, with little islands here and there of water lilies and the lush growth of bulrushes and arrowleaf, on which the muskrats, the wild ducks, and water birds feed unmolested

in the spring and autumn. Only a little while ago the ordered rich garden
of vegetables and flowers growing side by side was a dumping ground for
rubbish flooded with water from the silted-up channel of the brook; and
the fields were bare, poverty-stricken, covered in summer with a shabby
and ragged blanket of sickly green through which the yellow, pilfered soil
showed dimly.

In the house live Garth and May Forte. Garth is our indispensable
man. He drives the cars, does the marketing, gets the papers, waits on
table, meets arrivals at train and plane, and does a hundred other things
from time to time. May keeps house for us and cooks us delicious meals
when the stalwart Reba is off duty or on holiday. We could scarcely do
without them, but I respect them most because they *love* the old tavern-
house. They keep the lawn tenderly and put out flowers and have set up
boxes of flowering plants at the windows. One gets the impression that
the old house is proud again in its new dignity and happy because it is
loved, and because frequently people stop on the highway merely to
admire its simple dignified beauty, set among the immense and ancient
maples planted a century and a half ago by David Schrack.

I sometimes fancy that it looks at its own reflection, cool and clear and
beautiful, in the big fishpond, round and shaped like a mirror, that we
built across the road to contain the waters of the big spring and provide
fishing and swimming for all of us on Malabar and for our friends.
Around the big barn and the lanes, where there was not a shrub or a tree
but only a junkyard of abandoned rusting machinery overgrown with
weeds, the trees we planted—the dogwoods, the willows, the poplars—
have grown high and green, and in the spring the forsythia, the wichuri-
ana roses gone wild, and the kolkwoltzia are reflected in the clear waters
of the pond. ... But the work is never wholly finished. Each year there are
things to do when one loves a place. Each year it comes a little nearer to
that blessed vision of the imagination which is known to every good gardener and farmer. Things must have time to grow. It is like the painting
of a picture.

But we have wandered a long way from the spring about which I started out to write. I have wandered a long way, but somehow the wandering
was necessary, since in nature and on any farm, no one thing is separated
from any other thing. They are all brought together in what, to the superficial, may seem a disorderly and careless pattern, but to an observant and

understanding mind becomes one of the most intricate and beautifully ordered patterns in the world, involving man and beast, trees and wild animals and birds, fertility and beauty.

As I have written, the constant and beautiful spring comes out of the cliffs and sandstone rock behind the old house and tavern. Just below the spot where it emerges, David Schrack, the pioneer, built a big spring-house, piping the water from the spring itself through old hand-made tiles into the structure where it falls in a cascade into troughs hewn out of solid blocks of sandstone weighing tons. There is a deep trough for holding the cans of milk and a shallower trough for crocks of butter and buttermilk and for keeping melons and vegetables. The water is icy cold, and after flowing through the troughs it emerges into a brooklet clogged with cress and bordered by wild flags among the roots of two great willows. The springhouse was and is in itself a thing of beauty, but in a day of refrigerators, of quick freezes, and rural electricity, it is a relic of the past, a kind of souvenir of a distant life, and when we took over the Bailey Place the spring was half-forgotten and was not really working at all. The boys used it for cooling soft drinks, beer, and watermelon on hot summer days in haymaking time, but otherwise it went unused. The water simply flowed down the hillside to be lost presently in the rusting old farm machinery overgrown with weeds. The last woman to live in the house before it came into our possession carried the water from the springhouse to the house itself.

Then gradually over the years the spring came to go to work for us. We ran a pipe for water into the house itself and another pipe to fill a water trough with cold clear water for the cattle pasturing in the nearby fields. Where the old rusting farm machinery had been, we excavated a small shallow pond, where the watercress quickly formed a thick, solid mat of green which produces cress even during the hard winters of our country; for the spring water is *living* water and does not freeze. And on the other side of the lane, the waters of the spring were diverted from the cress bed to make a large pond of nearly two acres, which serves us in many ways. It provides an unlimited supply of water in case of fire; it makes a fine place for swimming in summer and skating in winter, and it provides bass and bluegills for everyone on the farm, whether for daily eating or to store up in the quick freeze for winter use. And out of it we draw the water to irrigate in hot dry weather ten acres or more of rich vegetable and flower garden and all the plots where we experiment with soils and plants.

But that is still only a part of the story. The waters of the spring flow by gravity from the pond to the nearby barn where the cattle spend the long winter, and thence through the cleared channel to provide during summer cold water and clean wallows for the sows and pigs in three hog lots. All of that has been established for a long time, but it was only in later years that we found still another use for the spring, perhaps the most important of all. It also solved once and for all the problem of the swampy spot at the roadside where the old wooden trough and willow tree had stood in my childhood. Like all good solutions, it was a natural one, with the virtues of both usefulness and beauty. We constructed there a roadside market to end all roadside markets.

The obvious spot was on the roadside just below the spring. The old stone troughs of the original springhouse gave us the idea for the design of the roadside market. We already knew from long experience what the cold water of the spring could do for the freshness, the crispness, and the tenderness of vegetables brought in from the hot sun-drenched garden. In a short space of time, the vegetables left in the cold water of the old stone troughs seemed to change. The wilt from the hot sun on the salads gave way to an icy crispness. The cucumbers were quickly chilled through and the sun-ripened melons achieved a coldness that no refrigerator could possibly produce. The Pascal celery became so crisp that if dropped to the floor it shattered like glass. This was not the *dead* cold of the refrigerator but the *living* cold of the spring water, gushing out of the primeval rock.

And so we destroyed the weedy, untidy little pond which had always been a problem and we excavated the earth of the hillside down to the level of the Pleasant Valley Road and set up a pavilion that was built back into the hillside, constructing a room into the hillside, itself with a stone dry wall on either side, so that one could look through the pavilion and past it to the steep sloping hillside and rocky cress-choked rill that emerged from the old springhouse. Inside the pavilion we built a series of troughs at different levels and channeled the spring water from above so that it welled up in the center of the top trough and from there fell in a series of small waterfalls into the lower troughs, until in one last waterfall it disappeared again underground on its way to the cress bed, the pond, the barn, the gardens, and the hog lots.

The dry walls and the earth behind them were planted with ferns and moisture-loving plants ... the same plants which once grew around the old wooden horse trough and willow tree. The troughs are built from the multicolored sandstone taken from the neighboring hillside, so that the structure seems to be a part of the hill itself.

All of us had a part in the building and design of the pavilion, and it was built as I like building things, with a skillful building contractor who does not assume the pretensions of an architect but has more talent and even genius than most architects. We worked it all out, day after day, a bit at a time, on the spot, much as I imagine David Schrack must have built the beautiful old tavern and the original springhouse higher up the hillside. It wasn't designed in cold blood and built from an exact blueprint. It just grew there, and when it was finished it was *right*, as if like the huge cut-leaf maples, it had grown there.

When we cleared away the pond and brought the level down to that of the adjoining road, an unsuspected blessing occurred. In an almost dramatic fashion, the excavation revealed the stairway and stone steps of the old house standing high above the road leading down to the roadside and the pavilion. All in all, we had achieved what every *good* farmer seeks to achieve … the combination of beauty and utility.

I began to write this on the evening we turned the waters into the pavilion that is also a roadside stand and I am finishing it weeks later while we are still recovering from the shock of great and instant success. On that evening when we first turned on the waters and saw the clear spring water gushing up in the midst of the pavilion, something curious occurred.

A few of us went down from the Big House to turn on the water, and presently people seemed to come in from nowhere. Neighbors and friends and even passersby on the road stopped in. The small children, who had captured a great snapping turtle, brought it along in a tub, and presently all of them were engaged in a general water fight, one of the pleasantest of sports on a hot summer evening. Even Charlie Schrack, the great grandson of David, the Mighty Hunter (himself an old man), turned up to see the final and almost complete rehabilitation of the place his great-grandfather had wrested from the wilderness of marsh and forest. People lay sprawled upon the grass, drinking beer and soft drinks, talking to the music of the spring water, as if an eighteenth-century French picture had come to life. The party went on until long after midnight. Nobody had been invited. They simply appeared from everywhere after Ivan Bauer and Dwight Schumacher, the building contractors who contributed so much to the creation of the spot, had announced that the water was ready to be turned on.

In a way, the party has gone on continuously since that opening night, for the sight of the spring water running from trough to trough of mottled sandstone is irresistible. No longer do tired hot horses and thirsty travelers in coaches stop at the old wooden trough to refresh themselves. There are no more horses and the passersby today come in automobiles of every description, from rich Cadillacs to jalopies. But they stop there again as they did a century earlier, simply to drink and to watch the water. Neighbors working on tractors in the hot fields drive over to drink and splash their faces and heads in the cold water, and citizens from all the towns in the area, sick of drinking evil-smelling chlorine, come to the spring to take home jugs of water.

In the hot summer of 1954, when drought afflicted most of the Middle Western countryside, the cold spring, unfaltering in its flow, was like a miracle. When other vegetable gardens were burned up by the heat and in the cities there was no water for the lawns or even for washing cars, our spring kept the big gardens and the lawns green and lush and fresh as an oasis in the desert.

A friend who spent some years working for American oil companies in Arabia told me a curious story about a spring and fountain which supplied the houses of a whole Arabian village. For centuries the women of the town had carried the water from the fountain to their houses, but with the arrival of the bustling Americans, who sought to impose the latest comfort and plumbing upon all the rest of the world (whether it desired it or not), the waters of the spring were pumped into a great tank to flow by gravity into each house, and the fountain, where Moses and Abraham may have drunk, turned dry and dusty. After a little while great discontent occurred among the womenfolk, a discontent which presently reached such proportions that there was an actual threat of rebellion. The women did *not* want the running water in their houses; they wanted their fountain back, for it was at the fountain that they met and talked and gossiped and heard the news; the American plumbing had condemned them to loneliness and boredom. And so in the end, the expensive pumps and tank and plumbing of the progressive Americans were abandoned, the spring ran again through the ancient fountain, and the women carried away the water in goatskins or in pots on their heads and everyone was happy again.

I think there was perhaps more to the story than merely the loneliness and boredom of the women in that Arabian village. There is something beautiful in the sight and sound of running water, not from a tap or faucet, but out of the rock or from the kind of beautiful fountain which in Italy becomes doubly beautiful with the arrival of the hot dry summer. I have seen fountains in almost every part of the world and always there is that sense of beauty and peace and abundance, which the sound and sight of running water can bring to mankind. I remember especially a beautiful baroque fountain in the courtyard of a great house in Brazil and again a fountain at the foot of high red cliffs on the beach near Trivandrum on the Malabar Coast. Its waters were sacred (as indeed are the waters of all springs) and the natives came from all parts of the coast to visit it and drink and bathe in its waters. The Malabari are a beautiful people, and in the evening there was a constant procession of pilgrims climbing and descending the narrow path up the high red cliffs, to and from a beach which is actually composed of powdered moonstone and garnet.

And so our spring and fountain and market stand, here in the middle of Ohio have become, like fountains everywhere, a gathering place in the heat of the day and in the long blue evenings when neighbors and friends drift in to talk to the sound of the spring water falling from trough to trough among the fresh green vegetables and fruit.

The spring has not hurt the business of the roadside stand. From the very beginning the problem of David and Patrick and Mary Solomon, who run the enterprise, has been to keep pace with the demand. The vegetables are of fine quality, but the spring water has brought them a freshness and crispness which makes their mist-covered leaves apparently irresistible. The fountain that is also a vegetable market is probably there to stay, for it is solidly built out of the rock of the hillside itself, and even if the sad day ever comes, which it may well do, when what is now open countryside becomes blocks of houses and streets, the fountain will probably still remain, for its waters do not come from shallow depths but from far down in the ancient sandstone, and perhaps I shall have the kind of immortality of which I have written before many times and which seems to me the only kind worth having ... that I shall be remembered because I helped to rescue the old house from oblivion and built the pavilion and the fountain to house the spring in the beauty and dignity it deserves.

And perhaps generations from now people will still read the story of the building of that house by the Mighty Hunter David Schrack, for the story is engraved on the stone of the fountain, and so David Schrack will not be forgotten. It is not a lavish payment to the memory of the man who understood the beauty of the valley and built his proud house by the side of the hardworking spring, which more than a century later brings so much to our lives and the lives of neighbors and passersby and even the farm animals and the deer and the quail, the raccoon and the bright-colored pheasant that come in the morning and evening to drink from the pond. Even if the day comes when we go out of business and the crisp, fresh vegetables and fruit no longer fill the stone troughs among the waterfalls, the fountain will still be there at the roadside to refresh the worker and the passerby and to serve as a gathering place in the long summer evenings for neighbors and friends.

At one side there is a modest plaque with the inscription, "To the memory of Mary Appleton Bromfield, who also loved this valley and found here peace, happiness, and abundance." She would have liked nothing better as a monument than a fountain filled with the fruit and vegetables which represent the richness and beauty of Pleasant Valley. She came there out of a worldly life, and from a distant part of the country, but she came to love the valley and the land with the earthy, tenacious love of a peasant whose feet are in truth rooted in the soil. She would have liked nothing better as a monument than a fountain which was also a roadside market stand.

## HOW TO MAKE A PARADISE

URING 1953 THE HONORABLE CLIFFORD HOPE, CHAIRMAN OF THE
Agricultural Committee of the House of Representatives proposed a
bill, later passed by Congress, which in time may prove to be one of the
most important bills ever enacted in terms of the general welfare of the
nation and the preservation of its natural resources. The bill proposed
pilot or model developments of small watersheds throughout the nation,
developments in which the federal government would provide the aid of
its experts in the whole field of conservation and would share expense
with local communities in the costs of bringing about such a development. The local watershed areas need only apply to the Secretary of
Agriculture and, provided all the conditions are met, are enabled to take
up at once a watershed development which would deal with problems of
erosion, of flood waters, of siltation, and many other of the damaging factors which cost taxpayers and property owners throughout the nation
such enormous amounts of money each year.

Already several such pilot developments have been set under way and
the writer had the honor of being the speaker at the opening of the first
of these projects to be set up. This initial development was begun on a
tributary of the Missouri River, the most troublesome and expensive of
all the great American rivers in terms of siltation and flood damage. The
bill has a twofold importance in that (1) it is the first real federal effort,

Much of the material in this chapter has appeared before in the author's *Out of the Earth*. It
is repeated here only because of the immense importance of watershed development to
the whole of the nation.

coming out of Washington, to deal with all these problems at their source, instead of at the bottoms of great rivers, where these problems become annually more acute; (2) it marked the logical development and advance of the whole pattern of soil conservation, of which the Soil Conservation Districts, which have worked so well and so efficiently throughout the nation, were the first manifestation.

The truth is that the battle for a program to prevent the continued erosion of our soils by wind and by water has largely been won, a battle in which that great American citizen, Hugh Hammond Bennett, conceived the plan and carried out the strategy. I am aware that there will be many who will protest the statement that the battle for soil conservation has been won, but out of many thousands of miles of travel and thousands of contacts with farmers and local organizations, I am convinced the statement is true. It is not so long ago that the vast majority of American citizens had never heard of soil erosion or soil conservation, but the advances in public education in this field have been immense in the last fifteen years, and today it does not astonish me to find a New York taxicab driver, a Hollywood actress, or a big business executive manifesting great interest and a surprising amount of knowledge on this subject which is so important to the welfare of the U.S.[1]

But perhaps the most irresistible and powerful influence in the advances made to preserve our natural resources has been the operation of the relentless laws of economics. Today, the lumber interests and the paper pulp industry are the most ardent of conservationists in the field of forestry and spend great sums of money each year in education and in actual reforestation and general restoration of the great forests which once covered so much of the nation and which for a century were brutally despoiled. The writer, cynically perhaps, has always believed that the greatest force in any reform or change for the better is always firmly rooted in enlightened self-interest, and throughout years of work in the field has never endeavored to present any problem in any other terms. The

---

[1] Perhaps the most striking work of education in this field was accomplished by a remarkable organization known as the Friends of the Land and established about twenty years ago. It was formed spontaneously by a number of American citizens who understood the gravity of the general situation and for years carried on an active campaign through speeches, magazine articles, books, the formation of local chapters of equally good citizens, and through an excellent publication known as *The Land*. The organization also worked closely with the Soil Conservation Districts Association, the Audubon Society, the Izaak Walton League, the American Forestry Association, the Mississippi Valley Association, the American Conservation Foundation and many other organizations concerned with the preservation of the great natural wealth and recreational facilities of the nation.

great industries interested in timber came abruptly up against the brutal fact that if there were no more forests, there would be no more industries based upon timber.

A somewhat similar realization has come to the American farmer, not without some suffering and disaster. By now any intelligent farmer realizes that if he does not take care of his soil, if he does not stop the erosion, he will, sooner or later, be out of business. As I have suggested elsewhere in this book, the days of more and more free land to be despoiled and cast aside are all over, and the costs of farming, even in taxes, have become too great for the bad and careless farmer to survive them. While it is very difficult to discover any truly accurate facts regarding the number of farmers who finally fail because of their neglect of their land, it is possible to estimate with a fair degree of accuracy that every year one hundred to one hundred and fifty thousand American farmers are liquidated, not as in the Russian fashion by being shot or sent into exile, but simply by the irresistible workings of economics.

In most cases there is no need to shed tears concerning these liquidations. In most cases such farmers *hate* their land, their livestock, and sometimes their children because of the low living standards and the lack of spending money. In most cases, in such times as these when employment is at an all-time high, it is possible for these unfortunate farmers and their even more unfortunate families, to move into the nearest town or city and find work in a factory. Here they will earn more actual spending money in a week than some of them ever knew in a whole year on farms which each year produced less and less income. That spending money goes into buying beers or going to the movies or purchasing a new car, and so benefits the whole of the nation's economy and serves to increase employment. In most cases, unless their farms are abandoned altogether or put to trees, they fall into the hands of young men of energy and intelligence, frequently trained in 4-H and Future Farmer organizations, or into the hands of neighbors who are better farmers and frequently set about a whole rehabilitation program on the land which had been so badly treated.

This squeezing-out process is related inevitably to the agricultural revolution which has been going on for a long time and the general transition, occurring throughout the whole of the nation, from a pioneer economy and civilization into an economy and civilization of maturity.

The liquidation is inevitable, and as it progresses will diminish to a great degree the political pressures which bring about some of the more idiotic of price support programs, for these political pressures come almost entirely from the bad farmer who cannot stand on his own feet and from the single cash crop farmer who puts all his eggs in one basket.

It was inevitable that the progression of the whole program for conserving our natural resources should begin with the individual farm, then progress into the stage of the Soil Conservation Districts, in which many farmers in a neighborhood or county practiced cooperation, and move finally into the development of the watershed unit embracing not only agricultural units but cities and towns as well. It was inevitable too that the whole program of controlling floods, erosion, and siltation should move beyond the antique and costly conception, coming largely from the Corps of Army Engineers, that the way to cure floods was by the construction of huge dams, which in a short time were destined to be silted up, and vast levees near the mouths of rivers, which were destined to break from time to time, and cost enormous sums of money to property owners, insurance companies, and taxpayers.

All of these forces operating together have brought about a new and more intelligent conception of conservation and the control of floods and siltation in the form of the new program for the development of whole watersheds upstream at the headwaters or in the middle areas drained by our great and frequently unruly rivers. Perhaps one of the best features of the Hope bill is that the project, which is in reality an experiment and a probing of the way, is set up upon a basis of cooperation among federal, state, and local governments, for, in solving all the problems involved, city taxpayers and property owners frequently suffer from damage far more than individual farmers. It is also true that many watersheds are not confined within the borders of one county or one soil conservation district or even one state. Essentially and primarily, watershed development and control is and must be the concern of the citizens of a given watershed area, although the benefits may extend all the way downstream to the Atlantic, the Pacific, and the Gulf of Mexico. Under such conditions it is apparent that such a development of small watersheds high up on most of our great rivers is of direct interest and benefit to *all* taxpayers.

While it is true that it is difficult or impossible to control floodwaters at the source in some areas such as the Rocky Mountains or the Badlands of South Dakota, it is possible to provide this control through small watershed development in the whole area of the South, the Middle West, and the Northeastern U.S.

In our own state of Ohio we have had perhaps more experience in the development of watershed controls than any other state in the Union, and the biggest project, involving the large basin of the Muskingum River, one of the Ohio's biggest tributaries, has long been established and serves as a remarkable example of the vast benefits of such a development when intelligently conceived and carried out. The Muskingum Conservancy and Flood Control District has created great interest not only throughout the U.S., but throughout the world.

In the year 1913, Ohio underwent a disaster which has never been forgotten by any of those old enough to remember it. One April morning it began to rain, not merely a good heavy rain, but a downpour described by the natives as "raining pitchforks and hay ladders." It rained all that day and the next and the next and the next, steadily and violently. This was no sudden but brief cloudburst such as the West and the Southwest sometimes experience. It resembled more closely the forty days and forty nights of rain described in the Old Testament.

For days the heavy rains continued without a break until even on the watershed of Ohio, in the highest part of the state, cities like my hometown of Mansfield were flooded. Trains on the transcontinental lines ceased running, powerhouses were flooded, and cities were left without light or power. Flooded sewage plants polluted the water supplies, and residents of cities were forced to drive into the country to springs for safe supplies of water. Bridges were washed out and farmhouses swept away by such currents as the oldest inhabitant could not remember, while towns and sections of towns were submerged up to second-story windows.

When the rains finally ceased and the waters had gone down a little, it was discovered that more than five hundred people had lost their lives and more than $300,000,000 of property damage had been done. This became known, and will remain known so long as there is a state of Ohio, as the Great Flood of '13.

Now, Ohio is a state unused to disasters. It knew periodically the flooded lowlands along the Ohio River and its tributaries. People had grown accustomed to them, but the citizens of Ohio had no experience with wild tornadoes or forest fires or earthquakes, or the periodically dis-

astrous floods familiar in some regions. They were resentful and indig-
nant that Ohio should be visited by such a catastrophe. It was an indignation of almost comic proportions, which James Thurber has described
in "The Day the Dam Broke," one of the finest pieces of humorous writing in American literature. It is a brief story about the Great Flood of '13
and of the bewilderment and indignation it aroused in the bosoms of the
citizens of Columbus, Ohio.

Fortunately, the indignation did not die away. The citizens, especially
in the areas which suffered most, decided that something must be done
about it. After a long waiting period in which there was much dissension,
something was done. The citizens of the area which had suffered most in
the valley and watershed of the Muskingum River, the biggest of its
instate rivers, finally drew up a plan known as the Muskingum Watershed
Conservancy District. It became a reality in 1933, and since then has operated to the benefit of those in the watershed itself.

The area was organized as a public corporation with the power to plan
and to construct and administer flood-control and conservation projects,
issue bonds, and levy assessments and taxes. It also possessed the power
to enter into contracts with the federal government or the State of Ohio
for cooperation in any project undertaken. Making decisions for the corporation were a board of common pleas judges and a board of elected
directors. Behind it all there stood the figure of one man by the name of
Bryce Browning, who, during the time of the Great Flood, had been
working with the Zanesville Chamber of Commerce and had witnessed
the death and damage in the Muskingum Valley. The plan was largely his
conception and it became his obsession. For nearly twenty years he
fought doggedly to keep alive the indignation of Ohioans and get them to
"do something about it."

Today Browning is secretary and director of the Muskingum
Watershed Conservancy District. He lives for it and probably will die for
it. His conception has created for the great Muskingum watershed and for
the people of Ohio a kind of paradise of recreation, and it has stopped
dead in its tracks the threat of future death and disaster from flood in an
area covering 8,038 square miles of rich and beautiful Ohio land.

It has created handsome forests and helped vastly to check all the
destructive erosion which was eating up rapidly some of the finest agricultural land in the world. It has created a necklace of twelve beautiful

lakes, as lovely as any to be found in the English Lake Country. There is boating, bathing, fishing, camping, and hunting close at hand for the Ohio millions who live in great industrial cities like Cleveland, Youngstown, Akron, Toledo, Columbus, Cincinnati, Dayton, and many smaller industrial communities. City dwellers need drive only a couple of hours at the most to find beautiful lakes and some of the finest fishing in the world. More and more former city dwellers and industrial workers have taken up small farms and holdings on the borders of the lakes and forests, where they live the year round, driving to their work in the towns and cities.

Ohio is, surprisingly, second in fur production among the states and territories, and much of its fur is contributed by the marshes and wild country of the Muskingum Water Conservancy District, where game, muskrats, and even mink abound. Each year more and more wild ducks, many of them of the deep-water variety, follow the flyway created by the twelve lakes, which join up, below the Ohio River, with the ladder made by the great lakes of the Tennessee Valley Authority area. There are no more threats of property damage and loss of life. But most striking of all, the whole district is the only area and project of its kind which for maintenance and continuance costs the taxpayers nothing at all. It supports itself and pays acre for acre the same taxes as the other land in the big watershed.

This all came about because the plan was conceived broadly, not merely as a power project or to prevent floods alone. Conservation of soil, water, and forests, increased property values, particularly in the cities, better agriculture, recreation, development of wildlife, and the planting of forests and rental of land for revenue were all taken into consideration. In short, a whole watershed through the center of one of the richest of our states, in the midst of the greatest industrial area in the world, was developed as an entity, with consideration being given for every aspect of its potentialities.

There is no longer any doubt about the workability and the great benefits of the pattern. The district has been in operation long enough to prove itself. In the spring of 1947, rains almost as heavy as those of the Great Flood of 1913 descended on the valley. Indeed, in the upper reaches the downpour equaled the violence of the 1913 flood rains, and not one cent of property damage occurred, nor was one life lost. The fourteen sluices of the great dams were closed and the floodwater impounded and held there, to be let out slowly over the succeeding weeks. There was no need for expensive levees, which in the past often broke and created more damage than if they had never existed. There still remained in each of the great reservoirs enough reserve space for floodwaters twice the size of the 1913 flood, a situation which is virtually inconceivable.

The interest of the federal government in such a project was undeniable, for the prevention of floods and siltation in the big Muskingum watershed was not only of immediate value to the area itself but also of great value to the areas bordering both the great Ohio River and the mighty Mississippi all the way to the Gulf of Mexico. The federal government provided an original fund of $22,500,000 for the construction of the upriver dams and on the Muskingum; the State of Ohio voted a fund of two million dollars for the acquisition of the necessary land. The operation of the dams and gates themselves was left in control of the Army Engineers, who still operate them.

The conservancy district itself works in close cooperation with both federal and state agencies having to do with all phases of flood control and conservation of natural resources. These include the state Division of Forestry, the federal Forest Service, the Soil Conservation Service, the State Commission of Natural Resources, the state and federal Agricultural Extension Services, the state Highway Department, the state Division of Wildlife, the Fish and Wildlife Service, the Geological Survey, and many private organizations having to do with all phases of conservation. On the borders of the district at Coshocton, Ohio, exists the largest hydrological and training-school station in the world dealing with the problems of water in all its phases. All these agencies contributed much to the establishment of the original plan of operation, and continue to make valuable contributions. Perhaps the most notable element in the operation of the area is the freedom from friction, feuds, and jealousies among all these agencies, perhaps because of the wisdom and the tact of Mr. Browning, the secretary and director, and the board of over-all directors. Since the unification and establishment of cooperation among all the agencies concerned with Ohio's natural resources, the arrangement has become increasingly more effective.

It is notable that the pattern and the idea are spreading within the borders of Ohio. On the Miami River, in southwestern Ohio, a series of dry flood-control dams already exists, although the basic plan offers nothing approaching the comprehensive developments of forestry and recreation provided by the Muskingum District. Both the Muskingum and the Miami Rivers are tributaries of the Ohio, but on the Maumee and the Cuyahoga Rivers, which flow into Lake Erie, plans are already underway for the establishment of watershed conservancy districts on the pattern of the Muskingum District. Similar plans are also under way for the Hocking River, a smaller tributary of the Ohio.

This development within the borders of the State of Ohio offers a solution, and a very nearly perfect one, for the troubles which afflict the watersheds of most states. Other states are beginning to follow the plans. Texas has set up in the Trinity River watershed a control plan extending far upstream to the headwaters which is as good for that area as a dozen gold mines or any number of new oil wells. The plan of the Muskingum District can be repeated almost everywhere by the citizens of the area themselves. The only exceptions are in sparsely populated wild areas such as those existing in the West and in the mountainous parts of the South.

In the past, taxpayers, property owners, and insurance companies have spent billions of dollars in repairing damages and in constructing and maintaining and repairing vast dams and levees at the mouths or halfway down our great rivers, when all the time it must have been easily evident even to a kindergarten child with a pile of sand and a watering can that one does not stop floods at the bottoms of rivers, but high up on their tributaries and in the forests and cow pastures. Despite this obvious fact, virtually the whole of our flood-prevention work has been concentrated until now upon huge and vulnerable dams and levees far downstream.

The inefficacy of such action has been proved again and again in our history. On the lower Mississippi, dikes and levees now carry the muddy waters as much as forty-five feet above the level of the surrounding territory and still that territory is constantly subject to the breaking and washing out of the expensive barriers and to the greater damage which occurs when a dam or levee breaks and releases water suddenly with ferocious violence.

Even now, the Army Engineers are planning and building dams for the Missouri Valley and other areas, and expensive levees and barriers on the lower reaches of that big, turbulent, and muddy river. As certainly as the sun rises tomorrow, those great dams will fill up with silt and become useless, and the great levees will break and release augmented destruction. It may be fun for the Engineers to build such dams and levees, and it may make a lot of money for the contractors on construction, and eventually on the highly profitable repair bills, but such tactics are immensely expensive to the already long-suffering taxpayer and for the property owners of the areas affected.

During the spring floods of 1947, the Muskingum Watershed Conservancy District dams proved completely the value and the efficacy of these upstream, tributary dams. Not only did the dams prevent the loss of a single life or a cent of property damage within the watershed area but they also impounded an immense volume of water, which did not add its burden to the already flood-devastated areas of the Ohio and the lower Mississippi. Damage on the Ohio and the Mississippi that year ran into many millions of dollars. The loss would have been doubled and more if the floodwaters of the Muskingum watershed had been added to the crest of the downstream torrents.

The operation of the Muskingum Flood Control District is really as simple as ABC. If, for example, similar flood-control districts existed on the upper tributaries of the whole Mississippi watershed, the floodwaters of that whole vast area could be impounded and released slowly when the flood dangers had passed or be used for irrigation or for creating electric power. The cost would be infinitely less than the construction of vast levees downstream, which periodically break, with great damage and loss of life, and of the huge dams which, after a few years, fill with silt and become useless. Moreover, the construction of these smaller dams upstream, very often in rugged territory, decreases enormously the condemnation and flooding of valuable agricultural land and the destruction of whole prosperous communities by the vast and ineffectual dams and lakes constructed downstream.

It is possible to imagine a Mississippi watershed, much greater in area than most of the countries in the world, which would in time be free of virtually all flood danger, with an immense saving to the nation in taxes and damages, and immense conservation of water for other purposes. Since the construction of the TVA dams and those in the Muskingum District, the floodwaters of the Ohio River, of which the Tennessee River and the Muskingum are both tributaries, have been cut in half. Likewise, the burden of floodwaters pouring into the lower Mississippi. Suppose the other tributaries of the Ohio—the Allegheny, the Monongahela, and a dozen smaller streams—were treated in the same fashion. The periodic flooding of the Ohio would be ended, and the flow of the Ohio, which lowers dangerously during the summer months, so that high-cost dams and locks have had to be constructed by the Army Engineers to permit navigation, would be greatly augmented and perhaps even stabilized as the floodwaters impounded by the dams were slowly released.

In time there would be created in the areas surrounding the upriver dams, areas for forests, fishing and hunting, boating, swimming, and other recreation, such as those which exist already in the Muskingum area. These, as in the case of the Muskingum Watershed Conservancy District, can be self-liquidating and eventually actually profitable, without any burden upon the taxpayer, or withdrawing any land from the tax rolls as federal- or state-owned property.

But the Muskingum Conservancy District has not only prevented damage, it has created and creates increasingly new wealth in many ways. To the small towns in and about the lakes and forests there has come a big revenue from tourists and from sportsmen who visit the area to fish and hunt. The reforestation program, which continues steadily, is creating millions of dollars' worth of new wealth and considerable employment. The large area of agricultural land owned, managed, and rented out by the conservancy to neighboring farmers is increasing steadily in production and yields. This is so because the agency leases the land on a yearly basis and exacts a program of the best agricultural practices. If the renter violates any of these practices, the lease is simply not renewed, and the land goes to a farmer who does a better job.

Of course, the forestry and agricultural programs reach much further than mere facts and figures. By the practice of a proper agriculture and the reforestation of bare, eroded slopes, the amount of floodwater coming from within the conservancy watershed area has been immensely reduced, and erosion and the consequent siltation of the lakes have been very nearly corrected. Within a very few more years, the runoff water and the erosion factor in the area will have been reduced virtually to zero.

The influence of such improvements extends, of course, far beyond the borders of the conservancy district, for their value is seen by the farmers of the whole big watershed. There is no doubt that agricultural and forestry methods throughout the greater part of Ohio are being greatly influenced for the better by the widening ripples of example and information coming from the conservancy district itself. Slowly the whole big watershed, in the past badly farmed, eroded, and with thousands of acres of land actually abandoned, is being reclaimed and transformed. As the water and topsoil are kept where they belong, even under the heaviest downpours, the flood damage, the siltation, and the necessity for using the dams for impounding water grow less each year. Except for the flood year of 1947, few of the dams have actually impounded water in any considerable quantity during the past five or six years. This follows the pattern which proves again beyond any argument whatever that floods and erosion are not stopped at the bottoms of rivers, but upstream in the fields, forests, and cow pastures.[2]

Some of the wildest and most beautiful country in the United States is to be found within the Muskingum Watershed Conservancy District. Some of this was once farmed-out, abandoned land which has been reforested, and some of it was simply wild, rough country; so wild that in certain areas it was completely inaccessible, except to an occasional fisherman, hunter, or woodsman. Much of this land still carries its virgin forest covering. It has now been opened up, discreetly, with good winding roads leading to canyons and ravines and inlets from the lakes where one could believe himself in the depths of wildest Canada or the Rockies, although he may be only an hour or less from some of the greatest industrial cities in the world.

Nothing is more satisfactory than in the evening to come upon an industrial worker who has knocked off work at four and brought the wife and kids out into a primeval wilderness for picnic supper and some fishing. Each evening the road which passes through our own Malabar Farm is bright until long after dark with the lights of the cars bearing fishermen and their families home after a long summer evening of fishing, boating, and swimming. Sometimes the car tows a small motorboat on a trailer, to be kept in the town garage until the next fishing expedition. Always the car contains a happy, healthy, sleepy family returning from an outing in some of the most beautiful country in the world.

[2]The results of improved agricultural practices in areas surrounding the Muskingum lakes and indeed the whole of the watershed, are extraordinary. Siltation records made at Pleasant Hill Dam and Lake, during the first year of its existence, indicated that at the existing rate of siltation, the lake would be silted up within seventy-five years. Tests made ten years later indicated an extension in the usefulness of the lake to three hundred years. Recent tests indicate that the siltation factor has been completely annihilated and that the life of the lake, as a lake rather than a silted-in swamp, has been extended indefinitely.

The sources of revenue in the district are many and varied. They come from long leases on cabin sites deep in the woods, from concessionaires who rent boats and fishing tackle or conduct sightseers through the area, from the rents coming in from leased agricultural land, from small admission charges for upkeep of special camping areas, and from the increasing revenue of the rapidly growing new forests. Within recent years the construction and rental of fine modern cabins on the lakes and in the forest areas has added greatly to the source of revenue. These cabins have electric heating and cooking equipment, beds, showers, and utensils. It is necessary to provide only bedding and food. Long before the beginning of the summer, they are all rented by vacationers. And perhaps the biggest chunk of income is derived from the leasing of fishing rights to the State of Ohio. In return, the conservancy has its own staff of fish and game experts, who practice the finest of fish and game management. Best of all, dance halls, honky-tonks, and saloons are prohibited throughout the whole of the area.

The sociological values are, of course, immense in terms of health and recreation and outdoor life, and in developing the kind of citizenship which the United States must have if it is to survive. Indeed, many workers and their families have set themselves up in cabins or small houses, and even on farms close to the conservancy area, and live there the year round, the wage earners driving back and forth to work.

All the following factors should be borne in mind: (1) The Muskingum Watershed Conservancy District has established a pattern, too little known, which could be applied to the upper watersheds of most rivers in the United States—a pattern which could save hundreds of millions of dollars a year in damage, many lives, and actually create new wealth in dollars and cents, and immense wealth in terms of health, recreation, and good citizenship. (2) It has withdrawn from the state and the nation no tax revenues. (3) It creates no permanent tax burden to be added to that which is already breaking the back of our American economy. As a matter of fact, the Muskingum Watershed Conservancy District actually pays taxes on every acre of land within its borders at the same rate as that imposed upon any citizen of Ohio. (4) It belongs to the citizens of the area and not to an artificially contrived bureaucracy in far-off Washington constantly at war with the citizens of the region.

The successful working model is there for anyone to see. It is the nation's best answer to those who believe that the federal government should own everything and that all government should come from a patriarchal bureaucracy located in Washington. I know of no better example of real working American democracy and its superiority over the National Socialist pattern.

I live on the edge of the whole beautiful area, which in summer is a kind of paradise. We rent one of its many farms. (It is one of five farms which make up our place, Malabar Farm; the four others we own.) From one of our own high hills we overlook several thousand acres of the conservancy district. I can testify that it is perhaps the greatest and most rewarding single project ever undertaken by the state.

Today, in the vast Missouri River Valley, there is veritable warfare over the fashion in which the flood and siltation problem shall be handled. The muddy Missouri offers the greatest problem of any watershed in the United States. For years it has carried billions of tons of silt along with its wildly destructive floodwaters down its lower course into the Mississippi and into the Gulf of Mexico.

For generations, and in many watersheds, billions of dollars of taxpayers' money has been spent in building downstream dams which only silted up, and levees which only broke and had to be repaired over and over again. Thousands of lives have been lost and billions of dollars' worth of property destroyed, all because the simple and perfectly visible evidence that floods are stopped upstream and not downstream has been persistently ignored by the Army Engineers, who have had charge of the flood prevention on our watersheds.

In the Missouri watershed, the same simple stupid pattern is being repeated to a large extent in the plans of the Army Engineers. Many of the planners advocate the construction of vast dams downstream on the various big tributaries of the Missouri and vastly expensive channels and "cut-offs" which will eventually either break out or fill with silt, and the same projects call for the condemnation and flooding of hundreds of thousands of acres of excellent agricultural land and the annihilation of whole prosperous communities. In the history of the world there have been few plans representing a more gigantic waste and futility. The object of flood and siltation control is not to build the biggest earthen dam in the world, not to construct the highest dam in the world, but to stop floods and siltation and conserve water, none of which is accomplished by the expensive and futile dams and levees built at the mouths of rivers.

It is reasonable to say that if all our watersheds were controlled as the Muskingum watershed is controlled, the flood and siltation bill for damages paid by the nation would be reduced by at least 75 percent, and there would be little need for the millions spent annually in repairs and dredging and other upkeep.

The pattern is there—a pattern which any child can understand. Perhaps it is too simple and obvious. More likely it is simply not grandiose enough or futile or expensive enough to merit the interest of most of our "Planners."

## OUR AGRICULTURAL ECONOMIC DILEMMA

E VERY MORNING THERE ARE SEVEN THOUSAND MORE PEOPLE IN THE
United States. Every time one tears a monthly sheet off a calendar,
there are two hundred and fifty thousand more. At the end of each year
there are two and one-half million more American citizens.

Meanwhile the amount of free rich virgin agricultural land to be
opened up throughout the nation has dwindled to nothing. There still
remains a little land that can be reclaimed and put to high production
through irrigation or by drainage, but the amount is infinitesimal in
comparison to the land already under grazing and cultivation. Beyond
this there is considerable eroded, worn-out, and marginal land which
*may* by a skillful agriculture be restored and brought back into production, but even this area is small by comparison with the total of the land
already in use.

At the present time and for some years past the diet of the American
people has been the richest and the best in the world in terms of proteins
and balance. The citizens of this country have available at any time of the
year an immense variety of foods of all kinds. American food is also of the
highest quality, the best inspected, and cleanest food in the world, and
strangest of all, it is the *cheapest* food in the world. Food in this country
consumes a smaller share of the family budget than in any other country.
It requires less work hours to purchase a loaf of bread, a pound of meat,

This chapter is the essence of an address given in 1954 before the New Jersey State Bankers'
Association at Trenton, New Jersey.

a bottle of milk, a pound of spinach, or what you will, and in the age of universal inflation this is the only standard by which costs and values of food can be accurately measured.

Not only is this food cheap but it is abundant. None but the *best* vegetables, the *best* broilers or roasting chickens, the *best* fruits, ever reach the markets, because the competition among growers is so intense that a grower would not attempt to take to market any but the best. The vegetable and fruit grower will tell you that it is no use and that he would only be compelled to cart his produce home again to be plowed under as organic fertilizer. Even in the canning industry, only top-quality food is used—never the rejects or second and third stuff as in most other countries.

In the midst of such abundance and quality, it seems odd that one should question its continuance indefinitely, yet there are signs that some day, not too far off, we may have to modify the abundant and excellent table to which Americans have been accustomed for so long. Most Americans, if they were faced each day with the monotonous and limited diet of even the average middle-class European family, would consider it a hardship.

The high-level diet of the American people can be continued even under a growing population pressure almost indefinitely or at least for several generations, but it cannot be continued without certain adjustments both in agricultural production and in agricultural economics as related to the rest of our economy. Part of the job rests on the shoulders of our farmers. The rest of it falls on the shoulders of the remaining citizenry, as does the choice of whether they have abundant milk and dairy products, and good cuts of meat twice a day (or three times if we count the breakfast sausage, ham, and bacon).

There is no question of starvation, or of anything approaching the limited and deficient cereal diet of the average Asiatic. It is a question of leveling off to the dietary standards of the average European city dweller and industrial worker which is a very long way indeed from our own.

With improvements in agriculture and a greater production per acre, the possibilities of raising food in quantity are almost without limit. Certainly under an agriculture as good as that of the top 10 percent of our farmers we could feed at least three to four times our present population at the high level to which we are accustomed. Dr. Firman Bear, of Rutgers

University, one of the world's great agronomists and agricultural authorities, has estimated that if we had such an agriculture and developed all
the possibilities of food production to scientific clam and oyster farming,
the production of millions of tons of fish in farm ponds and lakes, and
the utilization of algae for the production of high-protein food, we
should be able to feed somewhere near a billion population out of the
soil, agricultural, and marine resources we now possess. However, that is
the optimum, both in production and in what can be hoped for from the
farmer himself who is, after all, a member of the human race and subject
to all its fallibility.

The question we are concerned with is neither that of starvation nor
that of feeding a billion people in the near future, but with maintaining
abundance, quality, and reasonable prices for good food of all kinds. In
other words, maintaining a fine and even luxurious diet within the reach
of nearly all our population save for the absolute destitute, who will be
cared for in one way or another in any case.

The point toward which I am moving is the conviction that with a
rapidly rising population we cannot have such a diet without sustained
rewards for the farmer and the food producer in general.

There has been much loose talk in the war years about all farmers
becoming millionaires, but the facts are a long way from the illusions of
the city housewife, who, when she pays a dollar or a dollar-twenty-five a
pound for steak, fancies, without thinking, that this dollar or dollar-twenty-five all goes to the farmer. This, of course, is absurdly untrue. Today
out of every consumer's dollar the farmer gets an average of forty-two
cents. That means, of course, that for many products he receives much
less than that average. Not long ago in the celery market, the grower was
being paid four cents a bunch for celery which was sold to the consumer
on the retail market for from twenty-seven to thirty cents a bunch, while
the grower was being told by retailers and buyers that there was too much
celery on the market.

One hears much of the high prices the farmer is receiving for his products, but the fact remains that today on the whole he is getting less for his
produce on an average than he was receiving in 1940, while the cost of
everything he buys has increased, in some cases all the way to as much as
200 percent or more. The level of industrial wages and especially those of
the skilled worker has risen many times more rapidly than the prices

received for farm produce and these rises are in turn simply tacked on to everything the farmer buys, from chemical fertilizer through fencing to farm machinery. It is not merely the question of wage raises or increases in wages for one *kind* of worker. In the case of a piece of farm machinery, the very big raise in the wages of the coal miner is piled on top of the raises given to the steel worker, and after that come the wage raises of various skilled mechanics, assemblers, and finally salesmen and executives. On a piece of farm machinery the farmer pays all of these, while his own prices for many commodities have remained on the whole little better than stationary since 1940. (Various items such as beef have shown phenomenal rises but in the field of grains, vegetables, fruit, etc. there has been little change, and today prices paid for feeder cattle and lower-quality beef are actually below those of 1940.) Remember, I am speaking *not* of the prices paid *by* the consumer, but of the prices paid *to* the actual producer of food at its source.

It should be pointed out also that while the prices of the farmer rise and fall, the level of wages remains stationary or rises (barring a violent economic depression), and since 1940 the trend of wages has always been steadily upward in a ratio far out of line with the increase in food prices. Prices for farm products can and do fall at the present time as much as 25 percent or more in a couple of months, as occurred during the cattle market of 1952. These rises and falls are rarely reflected in the price to the consumer, because the wages and salaries, the profits of the middleman and the retailer, and other miscellaneous costs do not change. Often enough the retailer simply absorbs the farmer's loss as his own profit, and the price to the consumer remains unchanged. In the case of meat, the wages of the truck driver, the processing worker, the white collar clerks, the meat inspectors, the salesmen, the executives, the hauling rates either by truck or railroad, all remain stationary or increase. It is the farmer and cattlemen who take the whole beating. Rarely is the drop in the farmer's price for beef or hogs on the hoof reflected until months later on the retail market and then by only a small fraction of the loss the farmer has been forced to take. Frequently enough, as in the retail prices of beef, it is not reflected at all.

This constant rise in the wages of the truck driver, the processor, the white collar worker, the railroad worker, actually increases the cost of food without benefiting in any direct way the farmer who produces it, while at the same time and for similar reasons the cost of everything he buys creeps steadily upward. He benefits only in the vague and insignificant way that if all of these people have more money they can perhaps buy more of the food he raises. These raises in wages and costs all along the line, however, are passed on to the consumer, who is frequently the union worker himself, and the consumer yells, and assumes that the farmer is getting rich at his expense.

No area of farm economy or food production better illustrates this situation than the dairy industry, which I certainly know at first hand with a large milking herd. The price of milk to the city consumer is on the whole higher than it has ever been, but the price paid for raw milk to the dairy farmer is the lowest it has been in fifteen years. The rise in milk prices to the consumer has not gone to the farmer at all but mostly to the union labor engaged in processing and transporting and delivering the milk. Yet the average consumer and even the union worker's wife who buys milk thinks the dairy farmer is getting rich on the high prices she pays for milk.

It would be said honestly that the American farmer is least responsible of any element in our society for the increasing cost of food during the past few years. Even the American housewife who, when prosperous, wants only the best cuts and leaves a slack market for the poorer cuts, contributes to the high prices of the best cuts, which must carry the load of the lack of demand for the poorer cuts.

It is, you see, an intricate business. The farmer gets abused but is in reality scarcely responsible at all.

Cornell University estimates that behind every milking dairy cow in this country there is an investment of six thousand dollars without counting current labor and feed costs. For a herd of thirty milkers this would represent a total investment of approximately $180,000 on which the interest at 5 percent would be $9,000, yet there are milk truck drivers in the cities today who are making that much or more for one or two wholesale truck deliveries a day and without one cent of investment or risk of any kind.

I repeat that frequently the fall in prices paid to the farmer–producer is not reflected at all on the retail market. Recently during the rather aimless machinations of the price control boards, ceilings in reality kept the price of beef at the very top, actually above its natural price level because some retailers continued to sell at posted ceiling price levels although the decline in cost of food to them had already shown up vividly in their wholesale buying.

Such a condition, it is true, could exist only under conditions of heavy employment, high wages, cheap dollars, and a heavy demand for meat and usually for only the best cuts. In the end, when the price of meat on the hoof falls sharply, the farmer is forced to absorb virtually all and sometimes actually *all* of the loss with little or no benefit to the consumer. No one else along the line save the actual wholesaler or occasionally the retailer absorbs much of the decline in price.

Operations upon such an economic pattern are deeply unsettling to the steady production of foods of all kinds and tend to create gluts and shortages in the market, and this instability again works to the economic disadvantage of the farmer. When hog prices began to tumble in 1952, many hog farmers began to unload their hogs and even their breeding stock wholesale at falling prices. In such a fashion an "artificial" shortage is created within a few months. Prices rise, and many hog farmers buy back pigs for feeding and for breeding at prices much higher than those at which they sold their original stock. This is, of course, poor business perhaps, but the farmer cannot go on feeding hogs to sell them at prices out of all line with the costs of the grain he feeds them, whether he purchases it or raises it himself and measures it in terms of the greatly increased costs of his machinery, labor, gasoline, fertilizer, etc. Moreover, the dairy farmer producing clean grade A fluid milk for the city markets is forced to feed grains and supplements supported at a high level by government guarantees, while the livestock farmer's product goes unsupported.

This is, of course, leading us into a discussion of what is called "parity supports" in farm prices. This is a kind of mystical expression much in favor with politicians and is poorly understood, if understood at all by the average citizen and indeed by many farmers. Its meaning is really fairly simple. It simply means that the farmer is guaranteed a price for his produce in a fair and proper ratio to the level of industrial wages and the costs of everything which the farmer buys. The standard for this ratio was set by Congress arbitrarily as the ratio between farm prices and other prices during the year 1910, a year in which the farmer was receiving good prices and indeed had a margin of *net* profit actually larger than he has today. But behind this lies the iniquity which Bernard Baruch pointed out long ago during the war: no economy in which prices are controlled or subsidized only on certain commodities can work either efficiently or justly. The result can only be a mess, supportable only under booming inflationary conditions.

It is only fair and accurate to state flatly that for the past ten years, and especially at the present time, the farmer is *not* receiving prices commensurate with the parity ratio. If it were not for these parity supports, we should undoubtedly have less high-protein food than we have today and the prices in a day of abundant cheap money would be much higher than they are, and there would be actual shortages in many food commodities which we take for granted and can be produced today at reasonable prices. This is so because not all of our farmers, but only a comparatively small number of them, are top-notch farmers achieving high or maximum production per acre, per cow, or per unit of investment. About 10 percent of our farmers feed 50 percent of our population. The remaining 50 percent of the population is fed by about 30 percent of our farmers who are pretty good. Approximately 60 percent of our farmers, including the cotton, tobacco, fiber farmers, and the small suburban operators with a few acres listed as farmers by the Department of Agriculture, produce very little more food than they consume. The top 10 percent of farmers are so efficient in production and management that they can weather almost all vagaries of price and survive. The 30 percent who are pretty good can and frequently do cease to produce when farm prices show a sharp decline. The remaining 60 percent, and notably the food producers among them, would today all be in bankruptcy if there were no price supports. They simply could not make the grade. It is this element, plus the big, speculative operators in such commodities as wheat and cotton, who make the whole difficult business of price supports continually an acute political issue.

Some of the blame for the farmer's troubles must be borne by the farmer himself. Only 10 percent of our farmers could properly be called top-notch, efficient, productive farmers. As I have pointed out, about 30 percent are pretty good, but a fair 60 percent are not really efficient or productive at all. But do not forget that this 60 percent all have votes.

The same laws of production and profit apply to agriculture as to industry ... that the more you produce per dollar of investment, per man hour, per unit of machinery, or, in the case of the farmer, per acre, the less the product costs you and the bigger is the margin of profit, on the basis of a reasonable price. There are far too many farmers in the U.S. today who are farming five acres to produce what they should on one acre, and consequently whatever they produce costs them roughly five times as much to produce as it should—in terms of taxes and interest, labor, seed, gasoline, and other factors. Many a good farmer could increase his net profit from 10 percent and upward right on his own land through better production and greater efficiency, without asking any raise in price, but many of them don't and very few of them are likely to change. As I pointed out, they belong to the human race and share its imperfections.

However, the increasing value of agricultural land and the steadily mounting costs, of labor, gasoline, machinery, fertilizer, etc., are forcing countless bad and inefficient farmers out of business. The operations of economics are bringing about a rapid improvement in our agriculture because these farmers, who are actually being liquidated by economics, are largely being replaced by younger and better farmers and this tends in turn to make farmers as a whole less dependent upon programs of support. As nearly as I have been able to discover, the bad farmers are being liquidated at the rate of somewhere between 100,000 and 150,000 farmers a year. The whole battle for soil and water conservation has rapidly been won by economics; in the future the farmer who does not husband his land and maintain and increase its fertility is simply doomed by rising costs of everything that he buys.

In most cases these liquidations are not cause for tears. The liquidated farmers represent the last remnants of an obsolete frontier agriculture in which the farmer felt that his land owed him a living, and many of those being liquidated economically are farmers only by circumstance and not by choice. Many of them actually hate their land, their livestock, and even their families because their living standards are so low. In most cases—once they give up their farms—they go to town and get industrial jobs. They work five days a week, eight hours a day, and at the end of a week they frequently have more money in their pockets than they ever had before in their lives, sometimes more actual *spending money* in a week than they had previously in an entire year. And this money goes to buy commodities or to buy a beer or go to the movies, and generates both commerce and employment … which none of these men were able before to do while remaining on the land as poor tenant farmers or as small inefficient operators. So the whole

economic liquidation in agriculture is operating for the benefit of the
nation and its economy. This liquidating process, coupled with a growing
industrial development, lies primarily at the root of the immense increase
in the economic prosperity of the South ... once an almost wholly agricultural area lamely inhabited by incompetent, inefficient, and poverty-
stricken tenant farmers and dominated by the leech-like operations of the
big and greedy landowners and absentee landlords.

The point I wish to make here is that if all our farmers were as good
as the 10 percent who feed 50 percent of us, there would be less need for
any price support program.

I am aware that among my readers who have a casual acquaintance
with recent agricultural statistics, there may be some subject to rising
doubts about the trouble the farmer is in. I can only support my case by
two things: (1) that it is impossible to understand these things unless you
are actually *practicing* agriculture, not only on a serious but on an extensive basis; (2) by the statement, which is not new, that statistics can give an
amazingly false picture of any condition or situation, and many of the sta-
tistical reports coming out of Washington have done just this. They
remind me of one striking incident of how statistics may, while being out-
wardly correct, still create an impression of fantasy. It concerns a Maine
logging camp in which there were one hundred loggers and two female
cooks. One of the loggers married one of the cooks and the statistics
proved that 1 percent of the men had married 50 percent of the women.

In no field of our national activities is it so difficult to produce statistics of validity or accuracy as in agriculture and the livestock business.
You are dealing with a myriad of products, each one of them subject to
vagaries of the weather, disease, insect attack, flood, drought, or almost
anything you want to name. Neither the soil nor livestock are precision
machines which turn out parts or products each exactly alike, nor when
you start farming do you merely set up a complete factory and simply
start production by turning on the electric current. A farmer or livestock
man is dealing with every law of the universe ... and this can prove
extremely upsetting to statistics. If you would like to know how well the
farmer is doing today and whether he needs supports in order to make
any profit and so be able to purchase and consume the commodities
manufactured by industry and in turn provide profit for the manufacturer and employment for the working man, I would suggest that you go

today to the manufacturers of agricultural machinery or even of automobiles and ask them how much they are selling to the farmer. You might get a shock.

Moreover, there are scarcely two farmers in the nation who operate in a similar fashion, raise exactly the same crops, who have the same farm program, or even operate with similar degrees of efficiency. When you try to get reliable or significant statistics from which you can draw reasonable conclusions out of such a muddle, you are simply blowing up the chimney. There is only one way to know whether the farmer is really prosperous. Is he buying things and is he able to do so?

This nation rose to its vast power and strength during the industrial revolution. There were many factors which brought this about ... first and most important, the vast natural wealth of the nation in terms of raw materials and natural resources, and after that, the operation of competition, mechanical and inventive skills, the benefits of free enterprise, and the American system of capitalism, which is almost as different from that of Europe as it is different from Socialism. In this great industrial advance there arose a popular belief, not only in this country but in the world, that industrial development alone was responsible for our vast wealth and power. Agriculture came to be overlooked completely in the popular mind.

So it might be worthwhile bringing up, *not* statistics, but some straight facts which are scarcely known at all.

First is the fact that no other nation in the world knows such an abundance of food and agricultural commodities as this one and, more important, no other nation in the world has so nearly a perfect balance between agriculture and industry. Exports are vital to almost every other nation in the world and in most of them simply in order to *buy* and *import* food and raw materials. The export of our industrial commodities has never at any time totaled as much as 10 percent of our national production, and for some time past it has been considerably below that level. With us export is *not* a desperate necessity; we could survive quite well at standards possibly above those of most other countries simply upon the interior turnover and exchange of agricultural and industrial commodities within the borders of the nation. Export to us is a great convenience and is to some degree the margin which raises our living standards above those of all other nations, but it is not a *grim necessity*. We do not *have* to export in order to eat and survive as is the case with most European nations and many others.

This internal balance between production of agricultural commodities and the production of industrial commodities, and the ability of agriculture to purchase the products of industry and of industry and industrial workers to purchase and consume in turn the products of agriculture has been greatly underestimated in relation to our economy.

Another fact scarcely known at all is that the capital investment in agriculture—that is to say, land, buildings, machinery, and livestock—is greater than the whole of the investment in industry. And still another fact: as high as 50 percent and more of our population derives its income, wages, and purchasing power directly or indirectly from an agricultural base. This too may seem an extravagant statement but let us analyze it.

Agriculture is the whole base of economy in 95 percent of our villages and smaller towns. It is agriculture which buys the radios, the tires, the fertilizer, the insecticides, the farm machinery, etc., etc. It is also very largely the base of the economy in great cities such as Minneapolis, Omaha, Kansas City, Des Moines, Iowa or Montgomery, Alabama; or Memphis, Tennessee. In many of these cities the prosperity of agriculture and the livestock industry largely determines the prosperity of the insurance companies, the stockyards, the retail stores, even the value of real estate property. But let's go a step further. The whole of the vast agricultural machinery business, with its ramifications into steel, rubber tires, etc., is dependent entirely upon the ability of the farmer to buy. And beyond that there are the vast food industries whose employees are dependent upon agricultural production—the huge meat packing industry, General Foods, General Mills, and hundreds of smaller corporations. And a third or more of the steel business, the gas and oil business, the rubber business. You could carry on a listing of these facts (not statistics) almost indefinitely.

What does this all mean? It means simply that when over a long period the farmer cannot buy or if he chooses *not* to produce, or cannot afford to produce, men are out of work in our great industrial cities overnight … at first a few hundred and then a few thousand and then a few hundred thousand. It means also that all of these unemployed are fundamentally far less secure than the average farmer, who at least has a roof over his head and can feed his family. After a few weeks, these unemployed workers are on relief of some kind and, most disastrous of all, they can no longer buy in quantity the commodities produced not only by agriculture but even by the very industries which once employed them. The calamity pyramids, farm prices fall still lower, and presently the whole of our industrial economy collapses about us.

I think it fair to say that every depression in this country since the Civil War has begun when the farmer had not the money to go to town and buy a lantern, a set of harness, a radio, a tractor ... in short, when the farmer had no money to buy. Let us suppose that even during the past ten boom years, the purchasing power of the farmer has been suddenly withdrawn at any time. Not even the vast sums spent on war production could have cushioned the shock. Remember that among all the nations of the world our situation with regard to the balance between agriculture and industry is unique, and that in many ways a prosperous agriculture is the very foundation of our astounding economy and high living standards. This is merely a fact from which we cannot escape. If we had not this vast cushion of agricultural production and wealth we should be compelled to export great amounts of industrial commodities in order to feed ourselves and to maintain any sort of balanced economy. At our present high wage levels and living standards we cannot compete in overseas markets in many commodity fields, and in order to do so, wages would have to be leveled off by 50 percent or more and we would sink overnight to living standards and wages of crowded, resource-poor, overpopulated European nations.

I think you may have guessed by now the point or points which I am trying to make. Notably they are two:

(1) If this nation, with a constantly and rapidly growing population and a definite limitation upon its productive agricultural land is to continue to have the best diet and the highest living standard and the cheapest food in the world, there will have to be some sort of agricultural guarantee, security, insurance ... call it what you like. Like all other segments of our population engaged in the production of commodities, the farmer will not and cannot go on producing at a loss.

(2) If our industry is to prosper and industrial employment levels are to be maintained, again agriculture must have security and insurance in some form or other. I will go still further. I doubt that some form of security or agricultural price or insurance in some form or other will ever be

withdrawn from here on out forever. We have built our economy largely upon the fact that agricultural income *can buy* and that it *can* to a very large degree provide prosperity for the manufacturer and the banker and employment for the worker. And that it can produce the highest, the best, and the cheapest living standard in the world in terms of food.

There must be some degree of stability to the whole field of agricultural prices, and the farmer must be protected against erratic and calamitous falls in prices which he is called upon to absorb *almost entirely*, while wages, profits, and costs elsewhere than on the farm remain static or actually rise. There must be some recognition of the fact—nonexistent in any other field of productive activity—that everything the farmer sells, he sells wholesale, and that everything he buys, he buys retail; that he cannot, as can the industrialists (in many cases through tacit agreement or merely good business principles), come to an arrangement whereby certain commodities are tied more or less to a given and fixed price mutually agreed upon or determined by a uniform efficiency of production.

Agriculture is not two hundred or five hundred or five thousand large factories, operating in defiance of weather, of droughts, of floods, of a hundred obstacles, and disadvantages which in other areas of our society are virtually unknown. Agriculture is sixteen million farms, ranging in size from little five- or six-acre fruit and berry patches to the vast projects of the Imperial Valley or the huge cattle ranches of the Southwest. Raising crops year after year is not striking oil or merely building a factory and turning on a switch. Food is the United States's biggest single industry. It is still the world's fundamental undertaking, without which the world could not exist. Good agriculture is the most difficult profession in the world, and the good farmer knows more about more things than any man in any other profession, because he is constantly dealing with the universe.

I am aware of an outcry from readers—"What about surpluses?"

The surplus situation has arisen very largely out of the faulty workings of the price support program of the past few years. In the case of wheat the program was set up during the war to increase production in order to feed our allies, and most of all to benefit the bad farmer and the suitcase wheat farmer who plowed up vast acreages in the West to make quick profits out of low-yielding wheat production ... sometimes producing as little as ten bushels to the acre. Aside from the fact that the utilization of this land, which should never have been plowed up at all, has cost us hundreds of millions, perhaps billions, in the blowing and destruction of good soil valuable as grazing land, the taxpayer's money went largely to line the pockets of men who were not farmers at all but merely speculators. There is no reason whatever to support prices of wheat merely to make it possible for these speculators and low producers to continue in business. The good farmer with high production per acre does not need these high support prices and many do not want them. A year or two of free-market operation in wheat would quickly clear out the speculators and bad farmers, and price levels would adjust themselves to a point where there would be profit for the sound farmer and surpluses would vanish overnight, with a small surplus in production or none at all. Actually, the taxpayer and sometimes the farmer is paying out money to keep in business men who should find some other honest means of livelihood.

The wheat picture is a very clear one, but in general principle it applies to all other fields in which abnormally high price supports have been maintained. The withdrawal of a high price support program on potatoes quickly solved that situation, and it would do so in many other fields. The overly high price supports in many fields actually operate to handicap and damage one group of farmers while benefiting another. Meat and clean fluid milk for city consumption have no price supports, *but* the prices of the grain which goes into the feeding of the animals which make the meat and milk *are* supported, with the result that dairy and beef feeding operations are forced to pay high prices for feed while the prices of the commodities they produce slip downward until the margin of ruin is reached. In virtually every case surpluses have been created by support prices, which make it possible for the speculator and the inefficient and unproductive farmer to remain in operation at a profit. Very largely the whole of the problem has been created, together with so many other evils, by war conditions which neither Congress nor our citizens are willing to abandon and recognize as no longer necessary, and as being actually inflationary and in the long run destructive.

But none of these things have in the broader picture much to do with the economic facts listed previously, nor with the fact that our population is growing at a stupendous rate, and that we still have the best, the cleanest, the widest choice, and the cheapest food in the world. The point here is that surpluses and complaints from taxpayers have arisen largely from the fact that we are attempting to maintain wartime price supports at a high level which is no longer necessary to production by any but a small group of inefficient agricultural producers and speculators who might well be sloughed off to the general benefit of both agriculture and our general economy.

Modifications in price supports need not be pushed to the point where *no* sort of insurance or guarantee exists which can protect the farmer against constantly rising costs of labor and of everything which he must buy in order to operate. In this respect the parity formula is a sound one and can never be done away with save at the greatest risk to our whole economy and to the gradual decline in quality and abundance of foods and a genuine rise in their costs so long as our population continues to rise rapidly.

I am only trying to say that we couldn't give up agricultural supports no matter how much we desired it simply because we cannot afford to … and that goes not only for the farmer but for everybody. We have built the whole of our impressive economy and high living and dietary standards to a high degree upon internal self-sufficiency, and the keystone of the structure is the balance between agriculture and industry. It's too late now to do without the keystone.

## ON BUILDING TOPSOILS TEN THOUSAND TIMES
## FASTER THAN NATURE

THE GOOD MEN OF THE SOIL CONSERVATION SERVICE, TO WHOM THIS nation owes so great a debt, are in the habit of saying that it takes nature ten thousand years to build an inch of good productive topsoil, and in this they are quite right; but the statement scarcely goes far enough; it does not point out that the bad and ignorant farmer can lose that inch of topsoil in a single overnight rain, or that man, with knowledge, intelligence, and energy, can create topsoil many thousands of times more rapidly than nature, and sometimes he can do a better job.

With nature, the job of making topsoil is painfully slow, for nature can only lay down a leaf, a blade of grass or a twig one at a time to be presently attacked by weather, by insects, by bacteria, by every sort of agency, to leave behind only the most minute of residues. Man, as he advanced in technology, in ingenuity, in material prosperity, and as he has unlocked and put together again pattern after pattern of natural law, has been able, by understanding the methods and laws of nature, to speed up the process of building soil almost beyond the powers of imagination and of belief. Yet in and above all this there is a powerful law constantly in operation ... that in panaceas, high-pressure chemical fertilizer programs, shortcuts and makeshifts in agriculture, there is no really permanent good or accomplishment. It is only through remaining and operating within the laws of a given and *natural* pattern that man accomplishes any real and enduring results.

As has been observed elsewhere in this book, the real definition of science is the discovery bit by bit—a fragment of knowledge here, and bit of evidence there—and their final reconstruction into a given pattern which may frequently alter the whole pattern of man's existence. These patterns interlock endlessly so that in the end there is no fact in the universe that is not related to another fact.

In no field of science does this process of discovery and the reconstruction of a pattern give answers so evident, indeed so obvious, as in the science of agriculture, which itself deals with the whole of the universe and all its laws, constantly, day in and day out, in terms of economics, of nutrition, of chemistry, of physics, of hydrology … indeed of what you will.

The ultimate aim of every good farmer, of every good agricultural worker and scientist is to create a living soil which will produce not only the food without which man cannot live, but the biggest yield and the finest quality in food. Without such soils, man cannot in the end continue living and eating on any level of decency in a world which continues to increase its population at a truly appalling rate. Some—the more intelligent and progressive—have pushed beyond the comparatively simple problem of producing great masses of low-quality bulk food merely to sustain life; they have gone deeper in an effort to produce the *kind* of food and the *kind* of nutrition which create healthy and vigorous plants, animals, and people, and people with intelligence who may in turn carry the perpetual struggle for knowledge and more knowledge still further, and with it the vitality which is able to act upon that knowledge and translate the knowledge into the reality.

The whole pattern of activity since the very beginning at Malabar Farm has been working to achieve these ends. Sometimes the workers have blundered, sometimes there have been disappointments, and sometimes the owner has spent more than he should on this or that experiment or field of research … more than he should upon an enterprise which is in effect an experiment station supported not by taxes or inherited income and endowments but by the yearly earnings of the proprietor and of the farm itself. Sometimes he has worked fourteen hours a day, indoors and out, seven days a week to keep both ends, the outgo and the income, in operation and in reasonable balance, but every minute of the fourteen hours has usually produced in the end excitement, stimulation, and the satisfaction of achievement.

In all the operations at Malabar, it became clear in the very beginning that the whole pattern of all future operations and success was inevitably founded not only upon the process of restoring the soils of eroded, tired, abandoned land, but actually creating a productive topsoil over the whole of nearly a thousand acres, on which there was little topsoil and on 75 percent of the acreage none at all. And even the so-called "topsoil" was tired; it might well have been said that it was merely subsoil a little darker in color than the soil that lay beneath it.

I can think of no more discouraging sight to a farmer than the hills of the Bailey Place and the Big Hill known for two generations in the township as "Poverty Knob." It was an area which had been corned out, farmed out, pastured out, sheeped out, eroded out, and abandoned; where there was scarcely an honest blade of good grass but only poverty grass, brome sedge, sumac, blackberry bushes, and goldenrod. All of that land could not feed five cows for one week of the year and the cows would have had poor picking, indeed. So wretched was the condition of those hills that even the forest trees, which so easily take over any abandoned field in Ohio, could get no foothold.

It should be remembered in reading this that sixteen years ago there was very little knowledge or information available concerning the rebuilding of such land and even less practice in rebuilding it. It might justly be said that the rehabilitation of American agriculture began only about a generation ago when there was no more new rich virgin land to be had from the government free and for the taking. The old pioneer pattern was about played out and with its passing millions of acres of land had slipped into abysmally low production and final abandonment. The progress in knowledge and experience made concerning soils, what they are and how to make them, during the past generation has been truly immense, far greater than the average farmer, to whom all this knowledge comes in bits and pieces from countless sources, ever realizes. And few city dwellers have any realization of how important these advances in science, knowledge, and practice have been to him in terms of nutrition, cheap excellent food, and in terms of his pocketbook. The benefits to him in the present and for his children and grandchildren in the future are almost immeasurable.

In the beginning at Malabar we had to feel our way in the business of building soil like a man in the dark. There was much fumbling and some mistakes, for there was little knowledge available, and sixteen years ago even "grass farming" was unheard of save in the case of a few wise farmers who through their own efforts and experience had discovered its immense values in bringing back to life and production soils which, one might well say, had "died." If there is any doubt about this fact in the mind of the reader, he has only to go back about twelve years to remember the tremendous sensation created by Edward Faulkner's book *Plowman's Folly*, which became a national bestseller, was read even by Hollywood actresses, and attained a total sale of hundreds of thousands of copies. Mr. Faulkner dared to hint that the moldboard or turning plow was not the best implement in the world and that there were better principles of agriculture than those practiced by the vast majority of American farmers.

Mr. Faulkner, a neighbor of ours, was and is an enthusiast to the point of fanaticism, and consequently he was inclined, as are so many specialists, to believe that his own way of doing things was the *only* way and that it was efficacious under all circumstances, despite immense variations in general conditions. Later he wrote a second book called *A Second Look* in which he broadened his general outlook and point of view considerably and suggested that his own theories could well be both developed and modified in many ways in order to obtain the best results in the shortest length of time ... a factor of the utmost economic importance to the average farmer, and especially to the young farmer, who with a small amount of capital available is undertaking the colossal job of restoring fertility and production to his soils.

At the time Mr. Faulkner's book appeared, we were at Malabar already fumbling our way toward recognition of the same general principles he advocated ... principally toward the awkward inefficiency of the moldboard plow as an agricultural implement save under certain very specialized conditions. Out of a long experience in working with soils which began actually in my childhood, I knew well enough that there was no *one* way of doing everything in agriculture. Even today, after all our own experimentation and research and all the knowledge that has been developed everywhere, we do not treat two fields alike at Malabar. Each one must be treated differently, according to the steepness of the slope, whether the field faces north or south, whether it is a soil based upon gravel loam, or clay, or shale. We have even gone beyond the distinction between fields into a distinction of *parts* of fields and by doing so we get a steadily increasing yield and quality, and the use of our soils without any process of waste, destruction, or deterioration, but on the contrary of constant improvement.

Faulkner in his *Plowman's Folly* advocated the working into soils of all kinds, good or bad, of organic material to create a kind of mulch in which the green manures and even the roughest of materials decayed quickly and built good texture and increased production. He advocated the use of disks and field cultivators for the operation, but he made one error: he believed that it was only necessary to work the poor soils or the eroded subsoils to a shallow depth of not more than a couple of inches. This was true for the restoration of fields in which seedings were being made of grasses and legumes to be used for pasture or meadows only. Lime, some fertilizer, and any sort of rubbish chopped up together with the poor soils to a shallow depth provided a good seed bed and in time the grass and legumes would do the work of restoration, This was true, but it was a slow process, faster than that of nature, but still too slow to be economically sound. He did not anticipate the use of power and deep-tillage implements which have developed so rapidly among farsighted farmers.

In a sense, the methods upon which we have finally settled at Malabar for the building and maintenance of soils are in reality an immense extension and development of Faulkner's original idea. From merely scratching the surface and mixing in organic material to a depth of two inches, we extended the process to a depth of at least twelve inches and a maximum depth of about twenty-two inches. In other words, today at Malabar we convert the top twenty inches or more of soil into a kind of perpetual compost heap in which we grow constantly better and better crops *while* increasing the fertility and improving the texture of our soils.

In earlier books and articles, I have written of the process while we were in the process of getting results; by now we have been at it for so long and under so many different conditions and on so many different soils that we have no longer any doubts whatever of the efficacy of the system. The soils, their depth, their productivity, their texture are there for anyone to see, and where those soils exist there were only a few years ago merely bare, eroded subsoils, or a thin layer of dead topsoil which differed from the subsoil only in its darker color.

Put quite simply, the difference in tillage is simply that of mixing all residues, green and barnyard manures, chemical fertilizers, and mere rubbish *into* the soil rather than merely turning it over with a moldboard plow to be compacted into a narrow layer tightly compressed by the subsequent operations of heavy machinery on the surface. Under the old methods of "clean plowing" with a moldboard plow, natural decay and absorption of the materials become very nearly impossible and putrefaction is the result rather than the fermentation and the disintegration of organic material through the operation of bacteria, oxygen, and other elements which play a part in the natural reduction of waste materials into good and productive soil.

Any observant farmer knows that again and again when he plows a field, sometimes even two or three years after he has turned under with a moldboard plow the residues of a corn and cotton crop, he will simply turn up again the old fodder and cotton stalks which have never disintegrated into true soil at all but have merely been lying there mummified for months and sometimes years. In the contrasting process of *mixing* the materials deep into the soil, especially if it is reasonably good soil with a decent nitrogen content, it is difficult to find any trace of the crude materials which have been incorporated even after as short a period as three to four months. The residues, the manures, even the rubbish has been turned into good soil, with a consequent improvement of texture, organic content, and even drainage. There is little sense in plowing down green manures or corn fodder or even barnyard manures simply to bury them in a tight layer, to lie there perhaps inert and mummified for years without becoming incorporated in the soil itself as a part of its texture and fertility.

I have often made a wager that I would take half of a ten-acre field, fit it as we fit it at Malabar and, with all other conditions the same, get at least a 25 percent greater yield than any competitor could get from the other five acres fitted in the old-fashioned conventional methods in which the moldboard plow is used. But the greatest and most valuable results come not simply during a single year in which the mixing method is employed but in the accumulation of benefits which mount continuously over a long period of time, indeed forever so long as such methods are employed.

There are many reasons for this. Among them the principal ones are: (1) that all raw organic materials break down much more rapidly and are translated into soil texture, and a greatly increased availability to the plants both of natural fertility and of commercial fertilizer; (2) that in working the soils deeply with a ripping rather than a turning action and mixing the raw organic materials *into* the soil, a sort of sponge is created which absorbs in time nearly all rainfall and maintains the moisture and accelerates the capillary action of moisture moving upward from lower depths, instead of blocking the action by a *compacted layer* of material tightly compressed by the weight of heavy machinery; (3) that the ripping and mixing action avoids the creation of an artificial hardpan made at plow-sole depth by the downward-pressing action of the moldboard plow point (which is the only means by which the moldboard plow is kept in the ground); (4) that with the "ripping-mixing" method we are constantly farming the soil at twice or more the depth than is the case with the moldboard plow, and that in doing so we are deepening the good productive soil or at !east improving deep soils and subsoils which are already fairly fertile; (5) that by constantly destroying the old artificial moldboard hardpans and creating no new ones, the ripping action opens up the soil and permits the heaviest rainfall to penetrate deeply into the soil layer instead of forcing the rainwater to accumulate on the surface in shallow ponds, which sometimes do not disappear until evaporated by wind and sun; (6) that by aerating the soils deeply, the life of soil bacteria is fortified and their increase is greatly accelerated.

It seems to me that all of these factors should be self-evident. Certainly they must be so to the good small gardener who spades his own land and knows from experience that it is far better to mix and incorporate any manures or fertilizers he is using actually into the soil than merely to leave it in lumps and layers, in which condition it will do his crops little good and may even do them harm.

As the readers of earlier books about Malabar are already aware, we began feeling our way toward this system of deep sheet composting long before we found the proper tools to accomplish our purpose quickly and efficiently. In the beginning we used disks, but in order to get the proper mixing action and attain any real depth, the process forced us to go over the fields many times until it became a costly process in terms of time, labor, and gasoline. And the disk or the disk plow is not a completely satisfactory tool, for in time, if it is used perpetually at the same depth, it will create, less rapidly but just as surely, the same artificial hardpan as is created by the moldboard plow. We also experimented with a Scotch plow available from Canada, which was highly flexible and adjustable. By using this we managed to turn sods and even stubble on its side in the furrow rather than turning it over completely and burying and layering the raw organic materials. We then followed the Scotch plow with disks, which mixed soil and raw materials efficiently, although the mixing and fitting costs were still too high. But again the process was not satisfactory since it did not go deep enough and the Scotch plow, which is a form of moldboard plow, simply created the artificial hardpans we wanted to be rid of.

Throughout a large part of the time we were fumbling toward an efficient economic system of sheet composting, the tools for the process were actually on the farm. They are the so-called Graham plow and a giant rototillage machine known as the Seaman tiller.

Of course the Graham "plow" is not a plow at all if one thinks in terms of the old-fashioned moldboard plow and shallow tillage. It is actually a kind of giant spring-tooth harrow or field cultivator in which the tines are made of heavy tempered steel and so curved that, unless they are set at a fixed depth, they will pull themselves into the ground until they will stall a high-powered tractor in the attempt to bury themselves.

The Graham plow originated in Amarillo, Texas, where it was developed by a remarkable man named Bill Graham, who spent years and a fortune to get it into use. His "plow" was being developed at about the same time that water erosion and dust storms were rapidly destroying millions of acres of land in the great wheat country, and the moldboard plow was coming into disfavor because it turned over the soil and buried the straw and residues, leaving the surface bare to blow or be washed away. Further over-fitting simply created disaster. Countless wheat farmers had already begun to abandon the moldboard plow for the one-way disk, since this *mixed* the straws, weeds, and residues *into* the soil rather than burying them and so left the soil bound together when the wind blew, and open and porous to absorb water during heavy rainfalls. But the Graham plow did a much better and deeper job than the one-way disk in "binding" and mixing soil, and suddenly it caught hold and began to spread from Saskatchewan to the Gulf of Mexico.

Today it would be difficult to find a moldboard plow throughout the vast wheat country, and more and more the Graham plow has come to supplant the one-way disk. This revolution made a millionaire of Bill Graham, who through thick and thin had stuck to controlling the finances of his own invention, but in addition his name has been immortalized by the plow he developed, and to him the nation owes a debt for making a great contribution to modern, enlightened agriculture. Today very nearly every big manufacturer of agricultural machinery puts on the market some implement based upon the principle which Bill Graham developed and exploited in the face of every opposition.

The Seaman tiller, twin companion of the Graham plow, has had a somewhat similar history, and was largely developed in the beginning by Harry Seaman as a high-powered rototillage machine for building roads and airports. The tiller is supplied with a cylinder on which are mounted either steel knives or hooks according to the use to which the machine is put. This revolves at a high speed, chewing corn fodder or cotton stalks, heavy sods, weeds, and even underbrush into small pieces and incorporating these organic elements *into* the soil. When used properly it is one of the most valuable of really modern farm tools, but its use requires intelligence and some experience and knowledge. Where the farmer uses it as a quick shortcut implement rather than a sound agricultural tool, the results, as in the case of all agricultural shortcuts, can be damaging. That is to say, where it is used as a single complete fitting tool to prepare the soil for planting in a single operation and the soil is highly deficient in organic materials, the machine may break down the soil structure into fine particles and create a cement-like condition. However, we have never used it as a shortcut tool and never expect to, but have used it wisely with truly excellent results. It is one of the rules at Malabar that machines are not in themselves the solution to a perfect agriculture; they are as good as the intelligence of the man who uses them, and there is no implement in the whole category of farm machinery which cannot actually create losses and damage when used stupidly or carelessly.

The first and perhaps the most valuable use to which we put the Seaman tiller was in subduing the rubbishy vegetation which covered the whole of Poverty Knob. We had owned the machine for some time, but we had given it only superficial trials which had been unintelligent and without imagination. It was not until we acquired Poverty Knob, with all its rubbish, that we found for the machine its first effective use. We had attempted subduing the brush and weeds by disking with heavy disks, but the disks failed to "chew up" the weeds and small brush and to destroy the underground roots of difficult weeds such as goldenrod and blackberry.

At a very great expense in time, labor, and gasoline, we had managed to produce not a seed bed at all but merely a kind of mess in which, with the coming of another season, many of the perennial weeds, the saplings, and even the poverty grass would grow back again with the benefit of lime and fertilizer put on to get a pasture seeding. We had not driven the big tiller more than a hundred feet across the steep, rubbishy slope of Poverty Knob before we knew that we had the perfect machine for clearing and fitting such land. Once over was enough. Not only had the machine destroyed the unwanted low-quality vegetation, it had incorporated it into the soil and prepared a perfect bed for seeding. From then on the problem of clearing and restoring the weed- and sapling-grown abandoned land was solved.

This discovery of a single use of the Seaman tiller was the key to the rapid preparation of wasteland and the first step in restoring about three hundred acres of our own land, and eventually hundreds of acres on neighboring farms. With the machine as a base we evolved a formula of restoration which turned out to be cheap, quick, and effective. It was simply this—any time during the hot dry period of August through November we limed the field and then followed with the big tiller, which converted all rubbish into organic material and mixed it thoroughly together with the lime into a top shallow layer of the soil. The field was permitted to lie thus throughout the winter. No erosion occurred even on the steeper hillsides, for the rubbish, mixed with the earth, provided a perfect mulch, with no underlying artificial hardpan, which absorbed all rainfall and melted snow like a sponge. No drop of rain or melted snow had even a chance to start running away. In the spring the field was given one going over with a disk and seeded down to a mixture of pasture grasses.

Out of long experience at Malabar a seeding formula has been evolved which in our climate and soils makes the strongest and most enduring pasture. It consists of three pounds each of alfalfa, sweet clover, brome grass, orchard grass, Kentucky fescue 37, and one pound of ladino clover. Bluegrass is indigenous in our country and no seeding is ever necessary; once any degree of fertility has been established, it moves into all seedings, frequently with more vigor than we like.

This formula provides us with a mixture of grasses and legumes, all of them save the ladino and bluegrass deep-rooted, fast-growing, and drought-resistant. To be sure, most of the sweet clover goes out in the first year, but if allowed to produce seed on the stems left by the cattle, it will reseed itself year after year to some extent. If the pasture is rotated carefully, a certain amount of alfalfa can be maintained for three or four years. After that the pasture sod consists mostly of deep-rooted heavy-growing grasses together with ladino and the wild white clover which, like the bluegrass, comes in spontaneously.

In the first year of this restoration process, there is bound to be a heavy crop of weeds which come along with the grass and legume seeding. In our first attempts at restoration the weed growth was so heavy that there was only one solution—to clip the whole field with a mower once in late June and again in early August. The clipping turned out to be a great benefit to the seeding, for it provided a mulch over the seedlings which kept the ground cool and moist while the seedlings established sound and vigorous root systems. In the second year, the pastures were covered by a heavy sod and were virtually weedless. From then on their maintenance in good production was a matter of giving them a top dressing of manure or commercial fertilizer from time to time. About every seven or eight years they are given a thorough renovation by driving across them once with the Seaman tiller, which chews up the old sod, sometimes too thick and heavy for maximum production, into small pieces and at the same time makes an ideal seed bed in which the standard mixture is seeded again. As we have out of the thousand acres of Malabar more than two hundred acres of land too steep for the use of heavy machinery and the growing of crops, the Seaman tiller has provided the perfect means of seeding and maintaining first-class pasture at a minimum cost in the operation.

The Seaman tiller, used together with its twin, the Graham plow, provides the very base of all soil restoration and maintenance operations at Malabar. They are under some circumstances the only tools used in the operation of preparing fields for planting, and they fit together like a glove on a hand. They are especially valuable in the long rotations at Malabar where fields are kept for six and sometimes seven years in a heavy mixture of alfalfa, brome grass, and ladino clover for the production of hay and grass silage. In the course of this long period, the sod becomes extremely heavy and the roots of the alfalfa tough and as big around at the crown as a parsnip. In addition to this, the brome grass establishes very heavy, complex, root systems extending three or four feet downward. This is a condition in which the moldboard plow is extremely ineffective, even in the hands of a skilled operator. However, one needs only to run the Seaman tiller with knives on it once over the field at a shallow depth to chew the tough, heavy sod into small pieces. The Graham plow is then put into operation and rips up and mixes the truly huge tonnage of roots and green stuff thoroughly into the soil to a depth of fifteen inches and more. The result is simply a compost heap which absorbs all rainfall, provides easy, effective drainage, and within six months all roots and green material have been converted into excellent soil which under this process grows constantly richer, finer in texture and more productive.[1]

The Seaman tiller and the Graham plow operate perfectly on fields which are in the stubble of small grain or in standing corn fodder from which the corn has been harvested. In the case of stubble fields, the use of the tiller is unnecessary and the Graham plow works easily, mixing the old roots and stubble deep into the soil. In standing corn fodder, the tiller, equipped with knives, chews the fodder into small bits, none of them longer than two inches and less than an inch in width. The Graham plow, following along after the tiller, mixes this chewed-up material deep into the soil and within three to four months the corn fodder has disappeared,

[1]The whole of the ripping-mixing process is in truth a wholly new kind of agriculture concerning which little or no information was available in advance of our own experiments. For the sake of those who wish to experiment with the process, certain information and warnings are necessary to avoid troubles. One is that the farmer should never attempt to put the Graham plow into a sod field in which the sod has not been first chopped up by a rototiller or by heavy disks. Otherwise, the Graham plow will simply roll up the sod— and particularly a bluegrass sod—like rolling up a carpet and leave great piles and lumps of sod in the field which are extremely difficult to level off for planting. Occasionally a farmer using the Graham plow for the first time attempts to set it down the whole depth (approximately twenty-two inches) the first time over the field. The result will be merely to stall not only a big tractor but a small bulldozer. The proper system is to set the plow to the depth at which the tractor or bulldozer will pull it comfortably. (Usually this is six to eight inches.) After having gone once over the field at this depth, one can go over it a second time at right angles after setting the plow down another six to eight inches and even a third time diagonally if it is desired to practice really deep tillage, Although the two or three times over the field may seem expensive, actually it is not so in comparison with the moldboard plow, as the conventional Graham is more than twice as wide as a three-bottom plow. Also after two or three times over the field further fitting operations are less necessary than in the case of a field plowed by the moldboard. Very often the only fitting required after the Graham-tiller operation is once over with a disk or spring-tooth, and if the weather is dry, once over with a cultipacker or mulcher. I have gone into some detail regarding the system as it is so new that a whole new technique and approach to preparing fields is necessary.

turned into soil without a trace, improving texture, drainage, and providing great quantities of easily assimilable organic material.

The Graham plow, like the Seaman tiller, had been on Malabar for a long time before we worked out the proper invaluable use for it. Our first hint of its value came when we undertook operations with an ordinary Killifer chisel in an attempt to correct spots in some of the fields where the drainage was poor. All of these spots were on clay soil and some of them on fields on the very top of flat hills at Malabar. So poor was the drainage in some cases that the spots had actually become marshland in which marsh plants and grass were growing. Choosing the hot dry month of September when the poorly drained spots had dried out, we chiseled them thoroughly as deeply as we could go, and then we made a remarkable discovery. The poor drainage was not naturally poor; it had been created by an artificial hardpan made in the years before we acquired Malabar, by the downward-pressing action of the moldboard plow point at times when the clay was too wet for plowing. This tightly compressed hardpan came up in pieces two to three inches thick and as hard and impermeable to the passage of water as cement. It was a little like plowing up an asphalt pavement. Beneath the hardpan the subsoil had a normal texture and but for the hardpan would have served as a normal medium of drainage and water absorption.

A little study revealed clearly enough what had happened in the past. Here and there in the field plow-sole hardpans had been created slowly by the moldboard plow and each year these spots drained more and more slowly. Each season other parts of the field dried out and were ready for fitting long before these wet spots, but the farmer had gone ahead and plowed the *whole* field, the wet spots along with the parts of the field which had dried out. The result was that in clay soil that grew less and less well drained each year, the hardpan from the pressing action of the plow grew more and more impacted until it became wholly impermeable to the passage of water, and that particular area turned into a swamp, and the farmer simply abandoned those particular spots and plowed around them.

I have kept pieces of this hardpan as an exhibit in my office, because it is typical of a condition which exists on thousands and perhaps hundreds of thousands of acres of otherwise good and productive farmland, especially where the soils are heavy soils.

The breaking up of this hardpan with a Killifer chisel obtained immediate results. With the hardpan broken, no water stood any longer on
these spots even during the winter months, and within a year or two the
marsh vegetation, where it was not plowed up for crops, completely disappeared and was replaced by a volunteer growth of bluegrass, white
clover, alsike, and timothy.[2]

Once we discovered the immensely valuable results of using the
Killifer it seemed only logical that the Graham plow (which is really nothing more than a multiple chisel) was the tool to break up hardpans everywhere on the farm and prevent them from ever being created again. The
mixing–ripping process which resulted in the sheet composting system
now in use at Malabar for several years was the inevitable next step after
the chiseling.

There are other practical and economic benefits from the system. A
field which has been Graham-plowed can be left bare throughout the
winter without danger of erosion even on steep slopes, especially if the

---

[2]On one occasion at Malabar we had a visit from an old and experienced and prosperous
corn farmer from the west of Ohio where there is a rich gumbo soil, very flat and very
heavy in structure. It developed that the old gentleman had come more than 150 miles
simply to compare notes on the Graham plow. His story was this ... that he had on his
250-acre farm a system of tile drainage which had cost him many thousands of dollars. So
thorough was the system that the tiles were a rod apart. Yet gradually over a period of
years, the drainage became worse and worse. Aside from the fact that putting in any more
tiling would be a bankrupting operation, such a procedure was a violation of common
sense. For two or three years he dug here and there in his fields attempting to find blocked
tiles, but he found none. Then in despair he attempted a chiseling operation and what he
discovered led him, like ourselves, into a new kind of tillage, indeed almost a new kind of
agriculture. On some fields he stopped using the moldboard plow and attempted ripping
up his heavy soils with a field cultivator, but the cultivator was not really heavy enough to
be efficient and when he learned of the Graham plow he purchased one and from then on
the way was clear. As the gentleman described it, "I found that I had a hardpan made by
the moldboard plow which simply shut off the water from the tiles. There was nothing
the matter with the tiles, but the surface water couldn't get to them. It was simply held by
the hardpan until it was evaporated by wind and water. It steadily grew worse and worse
until sometimes I couldn't plant corn and had to wait for the land to dry out to plant soybeans, and in a wet year in some fields I wasn't able to plant anything at all because of
poor drainage. Now I am able to get into any of my fields in reasonable time to plant
almost anything I like." Then with a twinkle, he added, "Of course, maybe I had been
working my land a little hard in corn and beans and ran out of organic material." In any
average spring, you can drive through the flattish, rich Middle Western states and see field
after field, well provided with tiling, where the water stands in shallow pools until it evap-
orates. In nearly every case the reason is depletion of organic material and plow-sole
hardpans. The situation develops a kind of vicious circle in which as the field grows a little more poorly drained each year, the farmer pushes ahead and plows the soil a little
more wet each year to get his crops in, thus increasing the impermeable hardpans every
time he plants a crop.

field has been plowed on the contour, for the plow, chiseling the soil deeply, creates a system of alternate ridges and canals which block the runoff of any water. And the Graham plow, working the soil very deeply, permits the rain and the melting snow, however heavy, to sink deep into the ground and become absorbed.

Thus we are able to prepare all our fields during the late summer and autumn months, when other work is slack, and permit them to lie over the winter during which the action of freezing and thawing does the fitting work, which has to be done on many farms which practice spring plowing by many different operations. In other words, when spring comes, nearly all the work done with two or three kinds of implements and in many operations where the land is plowed in the spring, has already been done for us by the frost action, and our only job is to level the field for planting. This means twice over the field at the most. The soil, broken down by frost action, is already in a condition to plant vegetable crops without further fitting.

At Malabar we hold the belief that the more you keep off the soil with heavy machinery, the better it is for the soil and all crop production. Under this system of fall and even winter plowing with the Graham plow, the number of times it is necessary to go over a field to prepare for planting is cut in half. The results, so far as soil texture and tilth are concerned, are far superior to those on fields plowed in the spring which have been gone over again and again with heavy machinery to put them into condition for planting.

Thus far this chapter has been concerned only with machinery and the mechanical operations of a new kind of agricultural approach to the soil, but there are other factors of equally great importance to soil building and maintenance. One concerns long rotations, the use of green and barnyard manures, *deep-rooted* grass and legumes, and the steady maintenance and increase of organic material and the constant improvement in soil textures.

The history of much of our valley and indeed of most of northeastern Ohio is the history of millions of other acres throughout the nation. In the beginning our country was forestland, and when the forest was cleared away, the land was put partly to livestock and pasture and partly to a kind of self-sufficient general farming. Presently as the country developed and the cities grew, there arose a great market for timothy hay for feeding to the horses that pulled the carriages, the delivery wagons, and the beer trucks of the growing cities. And so our valley took to producing timothy grass in great quantities, especially on the hill land. During two or three generations millions of tons of timothy hay went out of our area to the cities and for a time the farmers were prosperous and built handsome barns and houses. But since little or no fertilizer was used and timothy hay is an essentially shallow-rooted plant, the yields began to go down and all the hilltops began slowly to develop a condition like poor spots on knolls or a rolling field. Then came the automobile and the market for timothy hay in vast quantities vanished within a generation and the farmers, looking west to the corn-hog country, saw farmers getting rich, and so they themselves went into the corn and hog business.

The results were devastating, for in this hilly country crop after crop of corn, grown in rows up and down the hills, quickly left the slopes bare of topsoil. But perhaps the most important fact in the whole history of the area was that in all the generations since the clearing away of the forest, the land had never been worked more than a maximum depth of seven or eight inches, and no deep-rooted plants, with the exception of wheat and rye, had ever been grown. The surface of the soil had, in one sense, been merely scratched throughout all that time, until presently erosion from corn and other row crops carried off on many farms what soil there was left.

Behind all this also lay the superstition, which still survives in some backward agricultural areas, that one must not plow up the subsoil and mix it with the good topsoil inherited from the remote past. Some farmers actually believed in their ignorance that subsoils poisoned what little of the topsoil remained ... a strange conviction which probably arose from two factors: (1) that on many farms for generations no effort of any kind was made to restore organic material; (2) that where, by accident, the subsoil was turned over and mixed with what remained of a once decent virgin soil, the immediate result was to reduce yields, and since no organic material was plowed into the hard-worked eroding soils, this condition grew steadily worse. On one farm now incorporated in Malabar, the farmer before us had been plowing some spots in his fields only to a depth of about two inches because he was unwilling to turn up any subsoil.

Even further behind all this, of course, lay the evil belief of many an old-time farmer that his land owed him a living, and when it did not provide him with a living, he came to hate it. I once even heard the story of a half-mad old farmer who in the spring each year took to beating his land with a bull whip to make it produce.

Certain facts must be self-evident, even to a child: (1) All good topsoil is simply a mixture of the original subsoil with an abundance of decayed and decaying organic material. (2) If the original material out of which the topsoil was made was ill-balanced or deficient, there would be no good virgin soil in the first place, since the subsoil was the foundation of everything. (3) Where there is a really good and balanced subsoil, farms cannot become worn out—for there is always a minerally rich subsoil which, when mixed with quantities of raw organic material, can be drawn on almost indefinitely, and rapidly becomes good productive topsoil. In some cases, as in that of most forest soils, the topsoil and even the subsoil, when properly treated with lime and put to a grass-legume program, can be made better and more productive than the original virgin soil itself. (4) Where there exists a minerally rich, minerally well-balanced subsoil of considerable depth, the process of restoring such land to high production can be accomplished quickly, easily, and even cheaply under an intelligent program. Indeed, a really good program may create a *better* topsoil than existed at any time previously in the many areas.[3]

---

[3]During the very dry summer of 1954 we had a striking example of the benefits of this loose deep soil and the value of deep-root penetration possible under the sheet composting method. We planted corn at the same time that a neighbor put in his corn, in the field exactly across the road from our own field and only fifty feet away. Except for the fact that we had a higher organic content, the soil conditions were approximately the same, as were the applications of fertilizer. The real difference was that our field had been prepared under the fall plowing program with a Graham plow and the neighbor's field was prepared in the conventional way with moldboard spring plowing, and fitting that required going over the field several times with heavy machinery. We planted the corn very thickly on our side of the road, about six to seven inches apart in the row, so that actually we had approximately three times as many corn plants per acre. During the hot, extremely dry weather in August the neighbor's field simply dried up, with a poor yield and with very chaffy corn. On our side of the road the corn showed no effects whatever from the drought and remained green and growing far into September with a very heavy yield of well-filled ears. Pictures of the contrasting fields, fifty feet apart, are contained in the photographic section of this book. Later, after the two crops had been harvested, we dug down to explore the root growth. In the neighbor's field the corn roots had gone downward seven to eight inches until they hit the old plow-sole hardpan resting on the subsoil and had then spread outward along the hardpan surface. When the dry weather came there was no moisture whatever in the top ten inches or more of soil and the corn fired and gave up the struggle. On our side of the road, we found an extreme root growth going straight downward to a depth of nearly three feet into soil that remained cool and moist through the summer. One other factor contributed: the thick planting on our side of the road shut out from the soil nearly all sun six to eight weeks after planting and prevented the penetration into

One of the great agricultural advances of our time lies in the fact that we have learned exactly how to manufacture good soils and that we have learned the why and wherefore of many things which take place within the soil itself, something which in the past was left largely to a set of what might be called pragmatic superstitions, or practices which came into use simply because they *worked*.

But let us not flatter ourselves too much, for it is a fact that the greater bulk of this newly discovered knowledge has come into existence only within the last generation, and that practices founded upon this great wealth of information are by no means prevalent. Indeed, as I have suggested elsewhere, some of the textbooks actually in use and some of the rules taught are shockingly out of date and are being still taught, although recent discoveries in science and research have proven them not only unsound but in some cases actually destructive.

The use of deep-rooted legumes and grasses and the part they are able to play, not only in the restoration but in the actual creation of completely new and highly productive topsoils, has been set forth in detail in *Out of the Earth* and has been treated elsewhere in this book. What is important here is the fact that machinery alone, even the most modern sort, cannot itself rebuild worn, old soils or create new ones, and that the effectiveness of deep-rooted grasses and legumes in this field would be greatly reduced and retarded if *only* the old techniques based upon the moldboard plow and other fitting implements, some of which are almost obsolete, were employed. In our experience at Malabar the regeneration of soils and the actual production of new and highly productive topsoils have been greatly speeded up and the costs reduced by employing a combination of rugged grasses and legumes, *with* modern power and deep-tillage machinery. Both elements, the mechanical and the natural, were required, working together to achieve the result desired. This, quite outside the other benefits such as drainage, control of erosion both by wind and water, the checking of water losses, better soil texture, bacterial life, etc., etc.

The evidence of success is present in the fields of Malabar, not only in the actual production both in terms of quantity and of nutritional quality, but in the soil itself. In 1953 a group of agronomists, headed by the writer's old friend, George Scarseth, one of the most imaginative agricultural thinkers of our time, made a series of deep borings and tests in

the field of the hot drying winds, so that the factor of evaporation was greatly reduced, as was the temperature of the soil in which the roots were growing. Close planting of corn, of course, is feasible only in soils of high fertility.

many of Malabar's fields. The results astonished even those of us working for years at Malabar who thought we knew our fields.

The borings were made in fields where only sixteen years ago there was no topsoil whatever, and revealed a healthy *dark* topsoil to a depth of about eighteen inches. Even more interesting was the evidence of the "subsoil" beneath this top layer. The truth was, what lay underneath the topsoil was not dead subsoil at all, but a soil lighter in color than the top level but full of organic material and of an excellent texture which crumbled in the hand. Here in this lower level it was the deep-rooted grasses and legumes which had done all the work save perhaps for the aid brought them by the destruction of the old hardpans at plow-sole depth which permitted *all* the rainfall, under every condition, to penetrate deep into the ground rather than running off down the nearest stream as it had done in the days when there was no top layer of dark soil at all and a plow-sole hardpan had prevented the deep absorption of rainfall.

This lower level of loose soil, filled with roots both living and decayed, extended downward for nearly three feet before the organic content began to diminish and signs of a true subsoil devoid of organic material began to appear.

These facts meant that the roots of the plants growing on the surface had approximately four and a half feet of good and even excellent soil which could be explored by them without barrier or hindrance. It was repeated evidence of the fact that too many farmers believe that when they turn under a green manure, all or most of the benefit comes from the green stuff which is visible aboveground. In many cases the contribution made by the deep, heavy, and elaborate root growth of a plant is of more value than the green stems and foliage which appear aboveground and frequently enough have a content of up to 75 percent or more which is merely water. We have pretty well concluded that the roots of deep-rooted plants are the greatest of all soil builders, not only because of the organic content which they provide but because they serve as well to break up hard, intractable subsoils, provide penetration for moisture and oxygen, and in the case of deep-rooted legumes bring nitrogen into the lower levels of the soil.[4]

[4]It is interesting to observe that a large acreage of soils which had only a shallow covering of dead topsoils or none at all ten years or more ago are now used at Malabar for raising truck crops of the finest quality. The actual *rapid* restoration of these soils only began within the past ten years, and *after* we had learned or rather worked out the whole new principle of deep tillage and deep-rooted green manure plants. Many a layman or amateur overlooks the great function played by such common weeds as the dandelion, the

In the years of experiment and research in the whole field of soil building, we have reached another conclusion which some years ago would have been regarded as heresy by many a soil conservationist, including myself. It is this … that the whole of the pattern of terraces, contour plowing, strip farming, diversion ditches, etc. would be unnecessary, or at least a matter of choice, in a truly sound modern agriculture such as has been outlined at least in this chapter. We have found again and again that on steep slopes where the soils have a proper organic content and are fitted in the proper manner, a condition so sponge-like and capable of water absorption is established that the factor of runoff water and of erosion becomes very nearly negligible.

Erosion was by no means general, if it truly existed at all, on the virgin soils first put to cultivation by the pioneers. It only began to occur when the organic material gradually became exhausted and there was no longer either decayed or, what is more important, *decaying* organic materials mixed with the soils. The steepest hillsides of Brazil, put to cultivation for the first time, do not erode at all, for the decaying layers of vegetation mixed with the soils drink up even the torrential rains of the summer season. At Malabar, by the use of deep-rooted green manures and special modern machinery, we have managed in a comparatively short time to create conditions very similar to those found by the pioneers in the virgin soils of the U.S., or today in the virgin soils of Brazil.

We should never forget that in a cubic foot of highly productive living, well-balanced soil, every law of the universe is in operation. It is not merely a cubic foot of mixed "dead" minerals, or even a cubic foot of inert minerals and *dead* organic material of the sort which frequently gives the deceptive and misleading black color to many soils and in itself has become with time virtually carbonized. A living, productive cubic foot of soil is teeming with life, with millions of benevolent bacteria of many kinds, with living fungi and molds, with earthworms and every sort of minute insect life, all in the *process* of constantly reducing the content of dead organic materials and even minerals into a high state of availability to the living roots feeling their way down through this mass of life, which

bull thistle, and many kinds of docks. These are plants with extremely strong and deep root systems which benefit the soil in many ways. Unfortunately, most of them flourish best in soils which are already on a fairly high production level. There is an excellent book on the subject of the use of weeds in agriculture by Joseph Cocannover called *Weeds, Guardians of the Soil*, published by Devin-Adair. He is also the author of a new book dealing with many principles of really modern agriculture called *Farming With Nature*, published by the University of Oklahoma Press.

is in turn feeding off the dead and decaying organic raw materials of what has once itself been living. Within such a cubic foot of soil, the fundamental rule of all life on this planet is constantly in operation ... the law of birth, growth, death, decay, and rebirth. When this law is in full operation on farms of reasonable natural fertility, many of the problems of agriculture—problems of erosion and "artificial" drought, of water loss, of disease, and even attack by some insects—tend to disappear. It is that kind of soil we have tried to create over the whole of Malabar Farm and in one degree or another we have succeeded in all our fields.[5]

---

[5][Specialized information regarding the modern implements mentioned in this chapter can be obtained from the Graham Plow Co., Amarillo, Texas, and Seaman Motors, Milwaukee, Wisconsin.]

# THE ROADSIDE MARKET
## TO END ALL ROADSIDE MARKETS

THERE WAS NEVER ANY PLAN AT MALABAR TO GO INTO THE BUSINESS OF vegetables and fruits or of raising them for sale in the city. It just happened; in a way we were forced into the business by the increasing number of people, some of them from neighboring cities and towns at a great distance, who stopped at Malabar to buy the vegetables grown in the plots which supplied the people of the farm and where the research and experimentation was carried on. Many of these unsolicited customers were in the beginning casual visitors to the farm or came down through Pleasant Valley for the evening drive. Many of them stopped at Malabar to look at the vegetables and flowers, for the garden area is always open to any visitors who are interested.

It is an inviting spot, for the main plots lie in a flat area just below the pond and from the lane that runs along the top of the high dam the whole of the garden plots, with their lush and neat rows of vegetables, are visible. The beauty and quality of the vegetables raised by David Rimmer and Patrick Nutt spoke for themselves and year after year the number of those who wanted the vegetables increased.

There were two reasons for this ... the obvious quality of nearly everything grown there and the fact that we had been working for years to eliminate as far as possible all dusts and sprays. When the visitors asked questions, they found that no inorganic dusts and sprays and no arsenical ones were ever used and that the amount of vegetable poisons used, when necessary, had declined to less than 5 percent of the dusts and sprays that had been needed twelve years earlier.

For years we had simply given away or fed to the pigs or plowed in the surplus vegetables, but the increasing numbers of people who wanted to buy them not only interrupted our work (for we did not want to be disagreeable) but gave us the idea of perhaps organizing the whole thing and making it simpler not only for the buyers but for ourselves. And so we backed into the whole business of market gardening and a market stand, which has become perhaps the most profitable undertaking per acre of all the projects at Malabar Farm.

Across the road from us throughout the summer and on weekends in winter live our friends, Bill and Sara Solomon, who some years earlier had bought three small run-down, half-abandoned farms, put them together, and set out to restore them with the help and advice of all of us at Malabar. They have four children: Mary, who is almost a young woman, and two girls and a boy, Carol, Sara Lou and Bill, who are triplets. Bill had a small shed which he had moved out from a parking lot in Mansfield. With all of this we seemed to have a setup. The shed was moved to the edge of the road, a homemade sign with the legend MALABAR FARM QUALITY PRODUCTS was set up, David and Pat brought in the surplus vegetables, the Solomon kids set about selling them, and we were in business. Now when people wanted vegetables we did not have to interrupt our work to get them. We had only to send them across the road.

From the very beginning the business far exceeded all expectations. Despite the fact that the market stand was not located on a main highway but on the winding unnumbered back roads of our countryside, there were plenty of buyers and frequently we ran out of stock. Some of them lived in the summer cottages along the big lake, some came from neighboring villages and towns, and some from a considerable distance. The fact that Pleasant Valley is beautiful country did not hurt our business. It had long since become an "evening drive" for hundreds of citizens from neighboring towns. And presently housewives who found the quality of the vegetables good told other housewives and our business grew.

By the end of the summer it became evident that our business had far outgrown the little shed with its home-painted sign and at the close of the season, when the last celery and squash and potatoes had been sold, the inspiration came for the roadside market that was to be a beautiful and airy pavilion with a whole brook of fresh spring water flowing through it. During the winter all of us sat around the fire in the evenings and "had ideas." Then when the frost came out of the ground Ivan Bauer and Dwight Schumacher came into the picture and went to work, and by late June the new roadside stand was ready for business.

This time David and Pat thought they were well set to meet the demand. They had doubled and tripled the production of many vegetable crops and extended their activities far beyond the borders of the plots surrounded by the hedge of multiflora rose into the adjoining fields.

The grand opening which followed the party of the Turning on of the Waters, took place on the Fourth of July weekend, and before the week end was over we knew that we were in for the same trouble as the year before—with all the expansion we were going to run short and fail to meet the demand. During the whole weekend practically everyone on the farm was in the garden harvesting beets, carrots, radishes, rhubarb— whatever was ready or available. The market sold out its whole supply four or five times over, and at nightfall all of us fell into bed. It was evident that we were going to need a kind of storeroom and nearby warehouse where we could stock vegetables the night before and have them ready for the daytime demand.

So Dwight Schumacher, the building contractor, was called in again, and within a week constructed for us on the other side of the road from the stand a storehouse which is ingenious and serves its purpose brilliantly by providing excellent cool storage for vegetables and fruits both in summer and in winter. The storehouse is built partly underground into the slope of the hillside below the road and is constructed of concrete blocks, with a damp earthen floor. But the most efficient and wonderful thing about the storehouse is that we again utilized the hard-working spring. Inside, we constructed high wide troughs of concrete through which a whole stream of spring water flows winter and summer. It comes in from underground and flows out through wide shallow troughs on the outside, which serve for washing and cleaning the vegetables. Over the washing trough there is an arbor which one day will be overgrown with grapevines. Trumpet vines are planted to cover the roof and ampelopsis to cover the small amount of wall that is exposed to the sun. As the vegetables come in each evening from the garden they are washed and cleaned and stored inside, many of them in the running spring water. By morning and the time of opening for business they are chilled through in live spring water, crisp and fresh and glistening. As supplies on the stand run out we have only to carry the chilled fresh vegetables across the road from the storehouse to the stand. One of the advantages is that the live spring water flowing through what is in reality a kind of underground cavern not only provides constant moisture but maintains an even temperature and prevents freezing even in the coldest weather.

In addition to the attraction offered to many a vegetable buyer and gourmet by the great reduction in the use of dusts and sprays and in the cases of many vegetables by their elimination altogether, the market stand has a strict rule that nothing but the best is put on sale at the stand. The best is not necessarily the biggest fruit or the biggest vegetable, as those who know and like to eat have long since come to understand. It is the flavor, the tenderness, and for the nutritionist, the vitamin and mineral content which in turn go with good flavor and good nutritive qualities. There is no better example of all this than the difference in quality and flavor and texture of the great, showy apples of the Pacific Coast and California and those from the Northeastern states. It is the reason why the real gourmet rarely buys a Western apple and why many a luxury shop in New York City sells only Ohio, Virginia, New York State, and Michigan apples.

The prices at the Malabar market are sometimes lower but much more frequently higher than those of the city markets and the great chain stores, but we feel justified because of the care taken in raising the produce and the obvious and indisputable quality and freshness of those raised in the Malabar gardens. In a practical sense we have been able up to the present to sell everything that we have raised, and frequently have run low on the supply or exhausted it. There is no point in the country roadside market attempting to compete with the chain stores on a basis of price, and the chain store customer is quite frankly not the kind of customer we seek, but rather those who want freshness and high quality and are willing to pay for it. One of the boys on the farm has jokingly suggested that we put up a notice, "Chain store trade not solicited."

One of the principal attractions contributed by the icy water of the big spring which runs both through market stand and storage house is that, by keeping the vegetables a few hours in or near it, they become chilled *through*, not with the dead cold of refrigeration, but with the damp live cold of the spring water. Once the vegetables are chilled through, it is possible to transport them on a hot day for a hundred miles or more in the trunk of a car without their losing their freshness. This factor has brought quality buyers from cities all over northern Ohio who come once or twice a week to load up their cars. In some ways, the market stand on an obscure and unnumbered back road in Ohio, has proven again, if proof were necessary, the saying of Emerson, that if you build a better mousetrap the world will beat a path to your door.

Nor do we follow the deceptive and even shady practice of many a roadside market in these times, where the proprietor buys his produce in the city market, hauls it out to his stand and then sells it to the casual passerby as "homegrown." Much of the stuff sold on the roadside markets today as "homegrown" is frequently of much lower quality and freshness than that which can be bought in the city markets, even after its long passage through the heat and dust of the commission market.

We have also made a specialty and increasingly followed the policy of raising and marketing vegetables which cannot be found or are rare in the average city market, and are varieties which are no longer planted by the big market gardeners for the general market. We have been producing Bibb lettuce for years, long before it was known outside the most expensive and excellent restaurants frequented by the gourmet, and we experiment constantly with new varieties of vegetables which are sometimes superior to old ones. We have a whole following of customers who come to us to get okra and the tiny red and yellow tomatoes and Italian paste tomatoes and white pearl onions, and there are customers so addicted to the happy and healthy habit of watercress that they will come from a great distance simply to get it dark green and fresh out of the icy, swift-flowing spring water. Sometimes they will consume a whole bunch on the spot while engaged in conversation.

We also raise old-fashioned varieties of cantaloupes, many of them far superior in flavor and taste to the standard one variety that is about all one can find in the markets everywhere, not because of its excellence but because it ships well and can be picked green and will ripen on the way to the market or while it stands around the backroom of a city store. Our own cantaloupes are ripened in the sun, go unpicked until they are ripened, and are then chilled in spring water. The old varieties treated thus bring big premium prices, and the only difficulty is to meet the demand. And we have found that by grading our potatoes into three sizes—(1) the big outsize ones for baking, (2) the standard grade A size for general cooking, (3) and the tiny size for cooking as "new potatoes"—they bring a bigger price, and the smallest potatoes, the ones to be boiled with their skins on, sell best of all. Again, this practice appeals to the customers who want to eat well and have the finest quality rather than the lowest price.

Celery root or celeriac for salads, soups, and flavoring attracts a whole category of customers, as do the winter radishes popular with the many citizens of Central European background in our part of Ohio. These grow to an immense size, and the Bavarians and Czechs eat them while drinking beer. They are of three principal varieties: one a kind of crystal white, one pink, and one (the Black Spanish) white with a black skin. Buried or kept in a good root cellar, they provide fresh radishes for the table throughout the winter.

The old watercress bed below the road quickly turned out to be a gold mine, for it required throughout the year little more than the labor of picking. Nothing is more difficult to find in prime condition on the city market than watercress, for once taken out of the cold spring water it loves, the cress begins to deteriorate in quality very rapidly and to lose, as it rapidly wilts and turns yellow, its qualities both of flavor and nutrition. One of the reasons for the high price of watercress in the city markets arises from this fact of rapid deterioration, as the process creates a high percentage of spoilage and unmarketability. At Malabar it is cut from the pond, bunched, and immediately taken across the road where it is put back again almost immediately into the same cold spring water from which it came. Good watercress, or any watercress, demands clean, clear, cold running water. If the plants have a gravel bed rather than muck or silt, the quality and flavor will be better. The original beds at Malabar were the native wild cress, smaller in leaf but more peppery and flavorsome than the "improved" big-leaf varieties. We now have at Malabar a crossed variety with larger leaves which seems to have retained the original peppery flavor of the wild cress. This was accomplished by the simple expedient of scattering a little of the "improved" but rather flat variety in all the spring streams. The crossed variety has come to dominate.[1]

One of the factors which give the vegetables at Malabar a high quality

[1]Watercress, now one of the most popular salads and garnishes, is a native of America and was only introduced into Europe in 1840. Incidentally, nothing is more delicious than cream of watercress soup or watercress dipped in the juice of a chicken roasted and basted with quantities of butter mixed with chopped chives, parsley, and dill.

is the very rapid growth. The more quickly a vegetable grows to marketable size, the better its quality in terms of vitamins, flavor, tenderness, and nutrition. As we built up the soils in the experimental plots at Malabar with quantities of barnyard and green manures, and a good program of commercial fertilizers, the period of growth to maturity became increasingly smaller, but the great change came when we began experimenting with soluble fertilizers and a variety of formulas suited to given plants, all of which included some twelve minor elements. Eventually these formulas and the machinery for mixing them in the irrigation water were put on the market under the name of Fertileze, by a corporation set up in New Lexington, Ohio, under the name of Nutritional Concentrates, Inc. The use of these soluble fertilizers has great advantages from every point of view, which are discussed in considerable detail in an earlier chapter. Not only have they improved the general quality of market produce, but the rapid growth has given us the advantage of supplying the early home-grown market when the prices are high and the profits greatest.

The whole of the business is set up largely for the benefit of the young people—David Rimmer, Patrick Nutt, and the four Solomon youngsters—who work for a percentage of the gross, in addition to their salaries, in the case of David and Patrick. They buy their own seeds and provide their own fertilizers, while the farm provides the equipment for the operation and the gasoline or the natural gas for the irrigation equipment. If it is possible to judge by enthusiasm, all of them are having a fine time while they are making money.

The prospect of further expansion of the market lies ahead, for the boys have uncovered an excellent market at good hotels, restaurants, and even hospitals throughout the area. Chilled fresh vegetables can be delivered fresh each morning with the minimum loss of freshness, sweetness, and general flavor, and we are able to provide a great variety of things which are difficult to find in the general market. The vegetables have not been shipped through a commission warehouse or knocked about a central market for hours and sometimes days, but come fresh from the garden, chilled thoroughly by cold spring water.

But all of this is only a part of the general expansion which has spontaneously grown out of the experimental gardens and the spring. The stand itself has gradually become a modest agricultural center at which books on modern agriculture are sold, and also supplies of specialized fertilizers, which are difficult to obtain. Orders are taken for the multiflora rose hedges, which have contributed so much to the beauty of Malabar and to the propagation of wildlife and the cutting of the heavy expense concerned with building and maintaining wire fences. Agricultural pamphlets and information regarding new and modern machinery, fertilizers, and other industrial products are distributed without any charge, and the stand is glad to cooperate with any company engaged in the manufacture of agricultural supplies. And, of course, there are the gardens across the road and below the pond which are open to visitors at all times. This year we are erecting a small open building which can serve as a meeting and picnic spot for the organizations and visitors who come to Malabar in large numbers on Sundays. It is a practical building, which in winter will be used for the storing of machinery.

And before the young people there is a whole, almost unlimited prospect of food production. During the war, while the husband of my youngest daughter, Ellen, was in service, she began making homemade jams and jellies, which, even before the establishment of the roadside market, sold rapidly. In the second year business expanded to such a degree that she engaged three wives of neighboring farmers to help her. Then the war ended and she decided that jam and jelly making and marketing had become an overwhelming job and, being only twenty and adventurous, she took off with her husband to Brazil where they now operate the big *fazenda* described in an earlier chapter. But a neighboring corporation took over the making of the jams and jellies on agreed specifications, and these are now among the products sold at the roadside market.

As the hog project at Malabar continues to grow, plans are also growing for the production of home-cured hams and bacon, both smoked and unsmoked sausage, smoked turkey, and Cornish game chicken. Much of the ham and bacon produced today in the U.S. is not cured at all; it is merely dead meat embalmed by injection of various chemicals. It produces, among other things, the most abominable of all things, *wet* hams. The changing history of hams and bacons has followed the general pattern of food production and processing in the U.S. While an increasing number of processes are employed to produce convenient and cheap, quickly prepared food, the element of quality has gone steadily downward, and we have come to live more and more with all sorts of inorganic chemicals and to consume them in steadily increasing quantities.

We shall move into all the new activities as time permits and capital becomes available. I have always preferred to allow a business to grow spontaneously, rather than to start at the top and go rapidly downward.

And it is not impossible that the same food enterprise will go into cheesemaking and utilize the facilities of an already established milk production, and a whole farm program of fine pasture, silage, and hay. It is possible that we shall end up with a good-sized mail-order business on our hands, but the roadside market with the big spring flowing through it will always be the showcase and the trademark, and the aim will always be quality, for those happy people who like food and know how to eat.

While on the whole the quality of food in restaurants and hotels has been growing steadily better over the whole of the U.S. save for a few areas, the natural quality of foodstuffs and raw materials in terms of flavor and frequently of tenderness has been going downhill. While many of the plant breeders have made excellent contributions, commercially speaking, to the whole field of vegetable and fruit production by creating new varieties that will ship or freeze well or lie around for days without actually rotting, they have done little toward improving quality and flavor. The showiest and biggest vegetable or variety of vegetables is not always the best, and most certainly the one that ships the best and keeps the longest is never going to win a prize from the Society of Gourmets.

The new varieties which have been created and are grown because of their shipping or quick-freezing qualities have done much toward giving the housewife green or frozen green vegetables the year round, but in the long run there is nothing like a vegetable that comes straight from the garden into the pot. Nothing loses its quality so rapidly as peas and sweet corn, but *old* vegetables of any kind tend to lose freshness and sweetness as well as nutritional quality. A tomato picked sun-ripened from a garden is in reality another vegetable entirely from the tomato that has been picked green and ripened in a truck or the backroom of a warehouse. And what can be sadder and more tasteless than two-day-old sweet corn which has been lying about in the heat, or string beans that are wilted and have lost that peculiar flavor and aroma belonging to string beans, which marks the really fresh string bean from the sodden, wilted pods that have been lying about even for a few hours. Perhaps the most fraudulent and inedible of all are the "keeping varieties" of early strawberries which sometimes come into our areas from parts of the South. They have all the flavor and quality of an old fishing cork impregnated with five-and-dime store perfume. They will keep all right: even nature has a tough time breaking them down, but I will leave the eating to somebody else.

If I seem fanatic about these things, it is only because, like countless other Americans, I like good food and the older I get the more pleasure I get out of eating.

And the foundation of all good food is the raw materials from which it is made. In no field is the old saying truer, that you cannot make a silk purse from a sow's ear.

While many of the new and marvelous shortcuts in food may make American life easier and perhaps even less expensive, and provide greater variety all the year round, they are certainly not improving the quality of food in the average American home. It may help the office wife to slap a frozen dish into the oven and warm it up in fifteen minutes, but she and her tired husband are welcome to it. The complete quick-frozen meal may permit the bad housekeeper to play an extra rubber of bridge at the country club, but I do not want to share the meal with her. The supermarket may provide, cheaply, radishes that have been lying about in a polyethylene bag for three weeks and resemble a wilted and elderly turnip in flavor, but where is that crisp tang which just burns the tongue and makes of *fresh* radishes one of the great delicacies? And it may please the lazy housewife to buy radishes in a plastic bag, with no tops on them, but she overlooks the fact or maybe does not care, that you can tell the age of radishes by their leaves, which wilt and decay very rapidly. Removing them may be a convenience, but it also has for the shopkeeper the material advantage of being able to palm off on you radishes which have not seen a garden for three or four weeks.

I remember my grandfather once saying that "a man who did not like children, dogs, music, and good eating" had only half a soul. It has stuck firmly in mind ever since, and as a rule I have found his observation very nearly infallible. Nothing can put me off a woman more quickly than seeing her being "dainty" at the approach of a friendly dog, pushing him away as if he might infect her with leprosy. That is about all I want to know concerning her and sometimes concerning her unfortunate husband as well. Nor do I want around the kind of woman who "hates" cooking and slaps together makeshift meals. These are not flash opinions, but the result of nearly forty years of experience, much of it with women, of considerable variety and intensity. Certain qualities, pleasant or unpleasant, are frequently indications to the wise and experienced of rocks and shoals ahead.

But the Malabar market stand has not only furnished us with an increasing income and a growing pride, it has also provided us with a lot of experience and fun. For the young people and the children, the experience of running the roadside market provides the kind of education for life that can never be found in a university. In dealing with the customers at close and intimate range, they have already discovered the great variety of the so-called human race. They have had to deal with the ill-tempered and the vulgar, the new rich and the rude, as well as with the pleasant people who like to eat well and look on life with a sunny point of view. They have had to deal with the haggler-for-haggling's sake, the kind of woman who attempts to pick over and squeeze everything, the kind who is just looking around, and the pleasant kind who sits around talking sociably. They have come to know the woman in the big new Cadillac who haggles over the price because she is starving the family at home to keep up with the Joneses, and the family which comes up in an old jalopy and buys ten dollars' worth of stuff without ever asking the price, because they like to eat well. They have learned that kindliness and good manners are more characteristic of the farmer and the industrial worker than of the country-club middle class. And perhaps most of all they have been touched by the warmth, the taste for quality and good food of those who have foreign backgrounds, either immediately or a generation or two back ... the Middle Europeans, the Balkan and the Mediterranean peoples, and now and then a Frenchman. Almost all of these peoples like to cook and to eat well and are proud of their cooking, and very few of them are ever divorced. I think, indeed, that the young people are getting an education that may be more valuable to them in living than anything they can find almost anywhere else.

By nature I am a notably poor salesman, of the "take it or leave it" variety, and perhaps my inclinations have crept too strongly into our market business; nevertheless I find a fascination in hanging about the stand watching and listening to customers and occasionally making a sale. Usually I am dressed in ordinary farm clothes and go unrecognized by most of the customers who come from a distance, but unless there is a rush of business I stand aside and when questioned about prices, simply say, "I don't know. I just work here." But it is all a fascinating business, and the fascination is not confined to me alone, for most of the visitors and guests who stay at the Big House also succumb to it and sooner or later find themselves behind the piles of fresh vegetables, with the spring water falling in rills all around them. Among the amateur and anonymous saleswomen have been Russian ballerinas, actresses, the beloved wife of our Ohio governor, and a few very famous names known to most of the nation. Occasionally they are brought up sharply for their inefficiency by a tart customer, but the rebuke does not seem to dampen the enthusiasm, because the whole thing is fun, as is the work of any really happy man or woman.

Perhaps the greatest product of Malabar Farm over the years has been the fun and the excitement which we have gotten out of what we are doing. It is not *always* that the results were fun and not always have we been successful, but at least most of us have been doing what we wanted to do in life, and what better justification is there for living?

At the market stand there is a beautiful view across the valley and forest, and always in the ear is the loveliest of all sounds—that of cold running water. Friends come and go, and neighboring farmers and the boys on the farm stop by going to and from the hayfield for a big hunk of cold watermelon that has been floating in the spring water or a bottle of spring-chilled beer, or a long drink of spring water from the gushing column that floods the stand. And there are seats where customers and neighbors and friends sit and gossip, always with the singing music of the spring as an accompaniment, and overhead the shade of the immense and ancient willows and maples, and just across the road the great bed of dark green, peppery, wild watercress that has turned out to be a gold mine for the market.

I am glad we went into the roadside market business. I have a feeling that the market will be there, run by the young folks, beside the "commodious mansion" built by David Schrack, long after I have gone. I am setting it up that way, and maybe never again will the big house and beautiful spring of David Schrack go unloved. And always, I am sure, there will be people who like to eat well and will have the best, or none at all.

## HOW TO LIVE WITH WEATHER

"People are always complaining about the weather, but nobody ever
seems to do anything about it."

—Charles Dudley Warner

NEVER IS THE WEATHER RIGHT FOR THE FARMER. IF THERE IS PLENTY OF
rain for the corn, the hay lying in the field is ruined. If the weather
is dry, it is good for wheat and bad for the oats and pasture. If it is too cool
the corn suffers, and if it is too hot the oats will make poor, chaffy heads.
There is an old Middle Western saying which runs:

> Wet and warm, oats and corn;
> Cool and dry, wheat and rye.

Perhaps it would be more truthful to say that the weather is always
*both* right and wrong for the farmer and livestock man. The only farmer
for whom it could properly be said that the weather is either wholly right
or wholly wrong is the single cash-crop farmer, and in the minds of many,
there are serious doubts as to whether he should be called a farmer at all,
but rather a streamlined speculator.

The old saw at the head of the chapter has endured and become known
to almost everyone for the very good reason that, like all platitudes, it has
become a saying because it is true, and therefore persistent and indestructible. Although irrigation in some form or other has existed throughout

historic times, it is only recently that its use has become widespread as a means of coping with the weather and at times of producing even what might be called an artificial climate, in natural terms. At the same time there have been great advances achieved to deal not only with droughts, but with a superabundance of rain or unseasonable downpours which can damage certain crops. New methods of drainage have come into existence, from the drainage pits which have long been used in Europe to the methods evolved by the Soil Conservation Service for channeling the massive rainfall of the Gulf Coast off the almost dead flatlands which border the gulf. It is only in recent years that in this country we have begun literally to do something about the weather, or more exactly to evolve methods and machinery which can thwart or at least cope with the droughts and sometimes with the unwanted and unneeded torrential rains.

At Malabar it became long ago apparent that at times we had heavy losses arising from conditions of weather. This was especially true in the early years when, with the universal shortage of organic material, our fields suffered badly in time of drought and heat. The soils, more often than not, had a cement-like quality which permitted, or rather forced, the water from any heavy rainfall to run off the soil rather than penetrating it. In addition to this, the wind and sun together simply sucked out what moisture did penetrate to a shallow depth. The enormous increase in organic material and broken-down humus gradually served to correct such a condition.

At the same time there existed over much of the land, and especially in the heavy soil areas, an artificial hardpan created at a depth of seven or eight inches, by years of pressure from the downward thrust of the moldboard or turning plow. This hardpan served in several ways to check and limit agricultural production: (1) It prevented the deep penetration of the fine hairlike roots which were seeking deeper levels where the moisture remained. (2) It served to block capillary attraction, the process by which moisture is drawn from deeper levels toward the surface. (3) It created in winter and spring large areas which were badly drained and where frequently, even on flat hilltops, water stood throughout the winter in pools on the field, far into the fitting and even the growing season. The result was simply that these areas produced only very low yields or were wholly unsuited for any sensible agriculture.

Deep tillage with the Graham type plows broke up these hardpans and with "sheet composting" methods, by which an abundance of organic materials were mixed deeply into the soil, many of our field crops, and even corn, began to show an almost total immunity to shortages of rainfall and a high resistance to actual drought.

However, this was about as far as we could go in coping with drought through the proper treatment of the soils themselves, and although these methods have been perfectly effective throughout the dry years following the Second World War, we discovered, once we began truck gardening on a considerable scale, that for growing vegetables and flowers, the proper treatment of soils was not enough to cope with the plant demands on moisture during long, hot dry spells.

In the experimental plots, of course, where we were able to work intensively, we had already added one other factor of the greatest importance in conserving and maintaining moisture and cool soil temperatures, highly important to growing vegetables and flowers of every kind. This was the element of mulching crops with a large variety of materials. Under such a system every drop of rainfall, however violent, was absorbed and, with the soils protected by mulch from hot sun and winds, the moisture was retained. Nevertheless we discovered that additional moisture produced astonishing results in the quantity and quality of production, and especially in the quick growing of vegetables which enabled us to reap the high profits of the early markets.

Although we already employed overhead sprinkler irrigation for experimental purposes, it became evident almost immediately, when we entered the commercial vegetable and fruit business, that a great expansion of irrigation facilities was imperative and would give us an immense advantage over competing growers in quantity and quality as well as early marketing. One other factor of considerable importance was the use of this overhead irrigation in coping with both early and late frosts, which might otherwise have destroyed considerable plots of vegetables being produced at the beginning or the end of the season when the prices were highest. In the face of an impending killing frost it was necessary only to turn on the overhead sprinkler system to equalize the temperatures of the immediate areas and protect the threatened crops or plants. This was immensely important in our own area, where frequently an early killing frost will occur in September and be followed by weeks and at times even months of mild growing weather. In the spring an occasional late frost, arriving only for a night or two, can kill off whole crops that have already secured an excellent start. The use of the overhead irrigation system at such times has saved us on occasions hundreds and even thousands of dollars.

In 1949 we began, together with the men mentioned in an early chapter, the research and experimentation into the use of soluble fertilizers through the overhead sprinkler systems. The results of this have been largely explained in the earlier chapters on soils; suffice it to say that the benefits could only be described as astonishing, and even perhaps miraculous, and demonstrated a whole new and important use for overhead sprinkler irrigation, not only for vegetable and fruit growers, but for heavy crop growers in much drier areas where irrigation was necessary, and for dairymen and breeders of registered livestock who were seeking not only abundant heavy pastures, but pastures as well of high nutritional values, which produce fine prize-winning livestock.

By way of recapitulation I will repeat the great advantages provided by the use of soluble fertilizer in overhead sprinkler irrigation. They are: (1) The fact that it is possible to correct immediately and almost overnight any deficiencies of any element, down to the minor elements of which the plants or crops show signs. (2) Being in liquid soluble form, many of these elements are absorbed as quickly through the leaves of the plant as through the roots. (3) Being in liquid form, in the very water used for irrigation, these elements are available immediately to the roots of the plants as well as through the leaves. (4) It is possible through the sprinkler-soluble fertilizer system to pinpoint and correct deficiencies which can sometimes cost the vegetable and fruit grower thousands of dollars, or greatly reduce production and financial returns. (5) It is possible to provide given crops the specific fertilizer formulas best suited to them. (For example, such things as celery, lettuce, spinach et al. need large extra quantities of nitrogen, while such crops as potatoes and tomatoes have an insatiable appetite for potash.) Through the application of these special formulas from time to time throughout the season, the quantity and quality of production and the rapid maturing so essential to good vegetables and fruits can be brought about to the optimum degree. (6) It is possible under skillful operation to apply insecticides and fungicides through the irrigation water itself, evenly, efficiently, and with an almost invisible labor cost ... merely that of dumping the materials into the mixer attached to the irrigation system. (7) The rapid and efficient, balanced feeding of plants by this system tends to produce a vigor and a rapid growth which make them highly resistant to attack by disease and even by certain insects.

Of course, in general, the lowering of labor costs and the heightening of efficiency are enormously increased through this system. Thus we have become virtually independent of climate and are able to carry out an almost perfect control in plant feeding and moisture requirements, with the result that, regardless of weather, of pests or diseases, we are able to produce on good living soils an optimum crop *every* year with all the profits that come from getting vegetables and fruits onto the early market. In this case we have indeed been able to do something about the weather. But let us not forget that soils and organic material still form the base upon which any sound and profitable agriculture or horticulture must forever rest.

Although this intensive operation has been confined largely to our vegetable and fruit operations, it can also be applied and extended with the larger irrigation systems to larger field operations for crops and for livestock. We have done a good deal of work at Malabar along these lines and expect to do a great deal more. Our observations on such operations (on a large field scale) give us ample evidence that the irrigation of crops, pastures, and meadows, even in the usually fortunate Middle West, is a decidedly paying proposition, especially in areas of rich, high-value land where high production and intensive operation per acre are imperative if the farmer is to survive.

A good irrigation system is the best crop insurance that I know; it becomes increasingly evident that even in areas of good rainfall, such as most of the Middle West, there is in the average summer always insufficient moisture to produce maximum yields in quantity and quality. In really dry years there can be no argument about a good system of irrigation paying big dividends. Of course in the South, the trans-Mississippi areas of the Middle West, and in the Southwest, the value of adequate irrigation can frequently spell the difference between ruin and big profits. Out of our own experience at Malabar and in travels throughout the country it has become increasingly and indisputably evident that the greatest inhibiting factors in general agriculture are moisture, organic material, and nitrogen, factors which are frequently and closely interrelated.

Behind all of this, of course, lies the question of water and where we get it from. Fortunate indeed is the modern farm which has adequate supplies of water for livestock, and even more fortunate is the farm with plenty of water for irrigation. Water, indeed, for industrial, urban, and agricultural purposes has become one of the nation's most serious problems, and the seriousness is increasing rather than diminishing. In our own rich state of Ohio the water tables on an average have dropped fifty feet or more within a generation, and in some areas it is impossible for a town to take in one more industry using water or even one more family because of grave water shortages. And in the great Southwest and in southern California, many areas are being forced to make a choice among industry, population, and agriculture because of the limited supplies of precious water.

At Malabar we are among the blessed, for we have a small stream running through the very middle of the farm which supplies us with water through the hottest, driest summers; we are on the borders of a large lake, and everywhere we have springs which can serve as supplies to small reserve farm ponds for storing irrigation water. The interesting fact is that through a revolution in agriculture and forestry, not only at Malabar itself, but among the farmers throughout our small watershed area, the flow of the springs and the equalization of the water supply in our streams has been steadily increasing. The heavy floods in winter and spring or during a torrential summer cloudburst have virtually disappeared from the watershed, and the billions of gallons of precious water which once flowed uselessly away to create downstream disaster and end in the Gulf of Mexico no longer escape, but are largely impounded where the rain falls. This precious water now goes into the ground where it is stored up to feed springs and small streams, many of which formerly dried up during the hot months, throughout the summer.

For the farmers who are not so fortunate in natural conditions, there are many ways of conserving, not only normal field moisture, but actual flooding torrential rains. It is possible, in dug reservoirs in the flat country and through the construction of dams and farm ponds in rougher country, to impound, for livestock and irrigation purposes, truly vast amounts of water which otherwise might simply run away to no purpose whatever. Even on farms where there are modest springs as a source of water supply, dams and ponds can serve to impound the waste flow, which can be used throughout the hot summer months. One of the most encouraging factors in the improvement of our agriculture and our water supplies has been the rapid construction throughout the agricultural portion of the nation of hundreds of thousands of farm ponds, which can provide not only substantial supplies of precious water, but a dozen other benefits of varying kinds.

All of this, of course, extends much further into the wise provisions for the management of small watersheds, which serve not only to conserve water supplies locally, but greatly to check floods and siltation downstream. The amount of wasted water through laziness, inefficiency, and greed which has been lost in the dry Southwest areas is appalling. Largely speaking, one of the greatest abuses in our water management and in the destruction of potentially good soils has been the over-irrigation of large areas of Texas, New Mexico, Arizona, southern California, and even some of the irrigated areas of the Pacific Northwest. As has been pointed out elsewhere, the reserves of flowing artesian well water, (and even pumped water) in some of the dry areas are limited, and it is possible that they have been stored up by nature over a period of thousands of years, and it may require thousands of years to restore those reserves.

The need for water is certain in time to bring laws regulating the uses and amounts of water for irrigation into many states which are not properly in the really dry areas, and where such laws, until recently, have not been considered. Priority and amounts in use are certain to go to those farmers and landowners who are already using irrigation water from public streams and sources.

Thus far we have been concerned only with questions of drought and too little rainfall, but for certain areas and for certain crops there is also the problem at times of too much rainfall, or rainfalls which come at the wrong times. At Malabar Farm, which is still essentially a grass farm, producing great quantities of forage and particularly of fine-quality hay, we have long been acutely aware of the damage caused to us by rains coming at the wrong time, which either ruin or greatly reduce the carotene and vitamin value of our good hays. We have long been able to produce the finest kind of forage it is possible to grow ... a mixture of alfalfa, brome grass, and ladino clover, upon soils which give us a mineral content in the forage high above the average. The problem has not been the question of *growing* top-quality forage, but the question of *harvesting* and *curing* it. Here we are brought sharply face to face with weather, for even a light shower on hay that is nearly dried in the field can cause great damage to the nutritional values. A really soaking rain lasting several hours can reduce a field of top-quality hay to a total loss.

It was obvious from the beginning that some type of operation which could dry hay artificially was the solution to our problem in an area in which rainfall was abundant and came during the summertime upon an unpredictable schedule of very local thunderstorms, so local indeed that sometimes we have been making hay in the sunshine in one field and in another less than a quarter of a mile away there was at the same time a heavy downpour. But when we began investigating hay-drying operations and equipment, we ran into many barriers and objections. Frequently the equipment was far too expensive for consideration by any farmer who was not engaged commercially in the production of high-protein supplements and alfalfa leaf meal. Many of the operations involved elaborate heating systems which were not only expensive to purchase and operate, but produced very high temperatures destructive to some of the valuable vitamins. Many of the blower operations involving ducts and scattered outlets were impractical, clumsy, and more often than not, inefficient. Then there was the problem of how to get the green, undried or partially dried hay from the field into the drier. Entirely green hay run through a field chopper tended to become tightly impacted when hauled into the barn, inhibiting completely the passage of air that was pushed through it by means of a blower. If it was baled while still containing a high moisture content and a high degree of greenness, the operation was complicated, slow, and back-breaking; and so for a good many years we abandoned the idea of artificial mow drying and concentrated upon the quickest and most efficient field drying possible, so as to be able to cut the hay, dry it quickly, and get it into the barn before a rain or shower.

In this operation we developed a system which was and is, I think, about as efficient as the best I have ever seen anywhere. Machinery played a large part in the operation, and in particular an implement first put upon the market by the Harry Ferguson Company ... a hayrake which did not, like the old-fashioned side delivery rake, roll the hay into a kind of rope which with every turning and windrowing only grew tighter and more difficult to dry, but which acted as rake and hay tedder at the same time. In other words, the Ferguson rake tedded the hay gently into a long loose windrow raised high above the ground, which permitted the wind to circulate freely through it. (At this point we should never forget that wind is a far greater factor in quick drying and in evaporation of moisture than the sun.) We had the machine for quick drying, but any machine is only as good as the intelligence and trustworthiness of the operator, and we in time trained boys whose whole job during the summer was to windrow hay efficiently for quick drying. The process, which I can recommend to any farmer, was to begin the windrowing operations with a Ferguson hayrake the moment the first wilt had gone off the hay and then, if the field cut was a large one, to simply keep going around the field on a hot day turning the hay constantly, raising it off the ground and into the air where the wind could serve as the drying agent. The hay was then baled with a much higher moisture content and greenness than could be risked with chopped hay, but primarily it was the saving of quality we were after and not entirely a saving in labor. The bales dried out quickly enough in the open mows, and in general we were able to make high-quality, field-dried hay, sometimes with the feed values of alfalfa leaf meal, quickly and efficiently. In good midsummer weather we have frequently made twenty- to twenty-four-hour hay of the highest possible quality.

How do you determine the exact moisture content and the tightness of the bale? Frankly, I cannot tell you, because here again the machine is not infallible and is merely as good as the intelligence and experience of the man operating it. All sorts of conditions affect the process, from the height of the bales piled in the mow to the tightness of the bale itself. It is something which you learn by experience, and I can make good hay a great deal more accurately by feeling the hay and following the baler and hefting the bales than I can with any machine that registers moisture content. It depends on how the air feels against your skin, how hot the day is, and a dozen other things which a real farmer will understand. The principal fact is that we evolved a system of making the finest quality possible of field-dried hay. But we had still not solved wholly the problems of coping with the weather; despite the rapid field-drying process, the thunder- showers sometimes caught us out, and days of rain sometimes delayed haymaking until the hay was past its best nutritional point.

It was only in 1954 that we evolved a long considered hay drier which operated upon the principle of *pulling* hot summer air through the hay rather than attempting to *push* it through or employing expensive heating processes. The hay drier is described in an earlier chapter so there is no good use in describing again its functioning. Suffice it to say that it is a cheap and simple operation involving an extremely low building cost and cost of operation, and it produces for us hay which is the equivalent in greenness and quality of alfalfa leaf meal, at a 75 percent reduction in labor and the costs of operations in the field. More importantly it has, like the irrigation system, made us independent of the weather. We shall still continue to make some field-dried, baled hay, for usually we have a production in excess of our feeding needs and some hay must be sold, but for our own consumption needs in the future, and as we develop more driers, we shall produce consistently hay of the highest nutritional value at much less cost in time and labor than in the past, and be wholly independent of the weather.

Of course one of the greatest means of coping with weather which is bad for haymaking is the silo, and silos have long been used for the storage of high-quality green meadow forage at Malabar, particularly the first cutting hay made in June during rainy weather, when the hay itself is of ranker growth and has less feeding value than the second, third, and sometimes fourth cutting of the same fields. We have from time to time put livestock through the winter on grass silage alone, and with good results, but the values of good dried hay are indisputable, and we have long since settled for a combination feeding of silage and first-quality dried hay.

The fact is that top-quality dried alfalfa-brome-ladino hay is in itself a concentrate of the kind for which many farmers squander money to buy when they might well be producing it at a fraction of the cost upon their own land. Top-quality silage has certain juices and milk-producing elements which are perhaps not present even in the best cured hay. On the other hand, all silage contains a very high content of water, some of it nearly as high as the 70 percent content of pasture or green meadow grasses and legumes. In skillful field-drying operations, and especially in the process of drying used at Malabar in which no high temperatures are employed, good quality hay is simply first-class pasture, silage, or green meadow growth with the water removed. In other words, for an animal to secure the same amount of minerals, vitamins, and proteins from silage as from top-quality hay, the animal would be forced to eat from three to four times the weight in bulk. Indeed, it is not possible for us at Malabar to feed high-protein, well-cured hay in an unrestricted diet to any livestock, from beef to calves, without getting into trouble, because the dried hay concentrates are too high in proteins and carotene for the digestive apparatus of the animals. There is, of course, all the difference in the world between the content of poorly cured hay, and hay made of poor quality material, and well-cured hay made of high-protein grasses and legumes.

Suffice it to say that in the field of forage and of vegetables, fruits, and some other crops, we have made ourselves largely independent of the weather. We are always sure of a fine quality crop of high production, grown and processed under the best possible conditions, and with great reductions in the cost of labor, gasoline, machinery, and time. This, it seems to me, should be the aim of all truly modern agriculture, forgetting never that the texture, quality, balance of elements, and organic material in the soil constitute the foundation upon which all the rest of the structure is founded. In all of these operations we have attempted no panaceas or shortcuts, but simply augmented or speeded up the normal operations of nature, and all within the economic range of *any* farm operation.

# THE WHITE ROOM

*A Somewhat Serious and Personal Chapter*

IN THE LOW, RAMBLING GREAT HOUSE OF THE BIG FARM IN BRAZIL, there is a large white room. The walls, made of the bricks and mud that come out of the fields themselves, are whitewashed. The furniture, consisting only of a wardrobe, a bed, a single chair, and a work table, is all white. There is nothing at all to distract the mind or the eye, and so, within that room, there is always peace. One can think, one can read, one can write without distraction. When I am at Malabar-do-Brasil it is my room, and I think I love it better than any room I have ever known.

Yet the room has none of the sterile, barren peace of a monk's cell, for it is set in the midst of life. There is a long low window looking out across the little ornamental lake, the orchard, and the jungle-bordered river called Atabaia, and all around it are the sounds of life ... the noise of the farm tractors and of the truck coming back from the little town of Itatiba, having discharged its great load of okra or mangoes or tomatoes; the morose bawling of the Holstein bull from the other side of the great brick platform where they once dried the coffee, and the occasional crowing of a proud defiant rooster. The workers, who come from North Italy, which produces one of the most beautiful peoples in the world, go by the window to and from their noonday dinner. Dogs, which are everywhere, bark, and now and then a puppy, feeling his way toward experience, finds himself lost from his brothers and his mother Jenny and sets up a loud and mournful wail. And everywhere there are children, the healthiest children I have ever seen anywhere in the world … Tommy, Terry and Lana, Dominique, Chantal, Mario, Anna, Yolanda, and Pilar, and my own grandchildren Stevie and Robin. In the evenings, especially on Saturday when there is a *bal* in the big Almaxerifada (which is only a baroque Arab-Portuguese word for warehouse), there is far into the night the sound of singing, as one by one, or more often two by two, the Italians find their way home in the moonlight. No wonder there are children everywhere … Italian, American, Portuguese, Japanese, Spanish, French, German, and mixtures of all of these. Sometimes it is Marcellino you hear singing on his way back from tending his pigs in the jungle down by the river. Marcellino is a giant, solid black man whose ancestors came long ago on a slave ship straight from the Congo.

But the White Room is the White Room, and while I am at Fazenda Malabar-do-Brasil it is all mine—a place to read, to write, to think, in what should be a country foreign to me but which is not foreign at all, but in some ways like France, a country to which I always seem to be coming home. It is a green country; even the hills which are almost mountains and surround the whole farm like the walls of a fortress are green, sometimes with the eucalyptus plantations, sometimes with the deep green of coffee trees, sometimes with patches of the primitive, impenetrable jungle. This is no desert, but a country of richness and abundance, and above all of life. The earth itself is old and rich but the people are newcomers, still engaged in taming the wild rich wilderness to man's will.

I do not know why the room was kept all white … the walls, the furniture, the ceiling. Even my daughter, who lives in the long, complicated old house, does not know; it just turned out that way, and it has become for me a center of existence, bringing back to me at times the deep-rooted feeling that free will plays a much smaller part in our existence than we would like to believe, and that somehow these things are arranged by fate, or the stars, or by good friends who have gone on to some other world.

Someday I mean to write a book about the progress of a boy born of Protestant-Puritan, Anglo-Saxon middle-class American background from Ohio to this white room in Brazil, by way of France and India and the whole world. The story is not yet clear, but I fancy that it will be the story of escape from all the influences which produce a tormented life like that of André Gide, the story of the elimination of nonessentials and creeds and doctrines and Protestant brooding and materialism, in favor of direct living. Very likely it will be called simply *The White Room*, because in that room, far from the turbulent life of our frustrated century, there is time and peace to consider all of these things after one has passed the meridian of middle age.

For the White Room came into my existence at exactly the right time in my life, and perhaps in the right place, for in the midst of all the fierce activity, the experiments, the comings and goings of people in that other Malabar in the vigorous, temperate climate of the North, there can be no White Room for me. There can only be the fierce activity, the driving turmoil that affects nearly all Americans, and prevents us at times from really living at all, permitting us merely to go through life in one long and somewhat noisy procession, always animated but at times empty. Yet the White Room came out of that other Malabar Farm in the North, without which this second Malabar, deep in Brazil, would never have come into existence, and so otherwise there could never have been a White Room, a room the like of which all of us need in life when the first excitements of youth begin to become commonplace and pall.

I have in Rio a good friend. He is an engaging man of perhaps sixty-five, a North Italian with blond hair, now white, and brilliant blue eyes fringed by the fine lines of a man who has lived well and enjoyed life. We were friends almost on sight, for I think each of us divined immediately that we were sympathetic largely because we had rarely said no to life. He said to me once, "Isn't it remarkable how a healthy, intelligent man can live wildly and well up to a certain age and then, quite suddenly, settle down and begin to accomplish all sorts of things?" I think I knew what he meant, from experience.

I have always been a singularly fortunate fellow, and in addition to that I have always known pretty well what I wanted, sometimes by a blind, driving instinct, sometimes consciously and even shrewdly. Knowing what you want and sticking to it is one of the great forces of life, and often enough it is a force which proves almost irresistible in a world in which most people never know quite what they want or are forever changing their minds about the goal.

From the time I was a young man—for material success came to me early enough—I have always had enough money to do whatever I wanted to do, with usually a little left over, and always I have translated money into freedom and experience, which is perhaps the only real value money has to offer and the only capital which is indestructible. I came to know nearly all the world and the peoples in it, so that today I feel at home anywhere and find that one cannot classify peoples vertically according to race, or color, or creed, but only horizontally, with lines which cut across all of these things. The bad and the good exist side by side in layers in every nation and among every group of people.

I have had all the automobiles I desired, although I have not much interest in them save that they must work and get me from place to place, and I have had great houses filled with fine and valuable furniture, and most of the time the best and richest food. I have never been forced into that rat race known as "keeping up with the Joneses." I have had a happy family life and fine and satisfactory children, and sometimes a very fun filled life outside the family. And I have had friends, good friends, almost everywhere. I have always had a tremendous good health and more vitality than is needed by one man. And perhaps the most satisfactory of all, I have made my own material success out of my own work and so know what it is worth and what it is not worth.

I still have the health and vitality and the fine children and the good food and a lot of the other material things ... the houses, the automobiles, the furniture ... but I don't know today quite what to do with all the material possessions; they have come to bore me and to constitute a burden. A strange thing began to happen to me nearly a generation ago, something which I suspect happens to any sensible man, in one degree or another, sooner or later. I began to grow tired of all these things, and I began to be bored by the life which plenty of people envied. I began to search for something; of what it was I could not be sure, and I built still another big house and filled it with more fine furniture, and for the moment I enjoyed myself heartily, as any man with even a spark of creation in him must do in such conditions. For I was building not only a new house and a whole new landscape, but, without quite knowing it, a whole new career and a whole new life.

As I observed in the first chapter of this book, I had long been homesick for my own country in Ohio, and I came back to it only to set up the same pattern of life which I had known up until then, a pattern I enjoyed to the full and have never for a moment regretted, but one which, without my quite knowing it, had become so used and frayed that it no longer had any interest. But Malabar Farm itself—the fields, the streams, the animals—became in time the very center of my existence, as nothing else in life has ever done. It became a small world to which I began presently to sacrifice everything else, and the sacrifices of time and energy were by no means confined to that small world, but presently became more and more involved with agriculture, with science, with economics, with all, in fact, that has to do with the material welfare of man, and not infrequently with the welfare of his spirit as well.

I was aware that the two were not so widely and hopelessly separated as some of the more sentimental would have us believe. There are among us the mystics who would have us believe that man can live upon the murky, tepid waters of contemplation and egotism alone, and there are the brutal, including even some of our scientists, who believe the full, rich life is merely a matter of plumbing and automobiles and air conditioning systems. Neither a poet nor a philosopher nor a scientist ever did good work with an empty belly in a damp and frigid attic; only the psychopathic derive inspiration from discomfort and misery, and what they produce out of discomfort and misery is of dubious value to any but themselves and their fellow psychopaths. On the other hand, the man who places all his values upon the mechanical, the material, is a sad empty fellow indeed and scarcely merits the trouble of conversation.

Over the years I became aware dimly that the true pattern of man's satisfactory and happy existence on the earth in our times constituted a marriage of the two elements ... that the material and mechanical advances should serve *not* to enslave man's mind and spirit in a wallow of mediocrity and time-saving, but rather to free his mind and spirit and foster their growth and enrichment. That, indeed, still seems to me to be the great problem of our times, whether in the utterly graceless and materialist world of the Communists, or the bustling, practical, material world of American capitalism. The problem, infinitely vague and difficult, was, in my mind, how to bring the material advances and the things of the mind and the spirit together; to reconcile and coordinate them in an age when material things have come to be looked upon in the popular mind as civilization.

For a long time before I came, by accident or through some hidden design, to the White Room, I was aware that I had been driven by something much more powerful than my own will and spirit. There were times when I was actually driven to and even beyond the limits of a great health and vitality into efforts and acts which, by practical and material standards, seemed useless and perhaps even meaningless. Certainly there were, among my friends and acquaintances, some who believed that I was a little mad. They believed that I was merely squandering not only considerable amounts of hard-earned money, but time, energy, and vitality. What was this force that drove me? What had come over me? Where had it come from? I found that I myself at times had gloomy doubts.

I asked myself these questions because, despite a life of great activity, there have always been moments when I found it necessary to sit down for a time and ask myself whither I was bound and why, times when I needed to get my breath and have an accounting.

Saint Francis and Saint Augustine, after wild and even dissolute lives in their youth, offered the example of turning to religion, and spent the rest of their lives preaching or writing theology. Such a progression could never have occurred to me; it was not in my nature, and both careers seemed a bit old-fashioned and out of key with the realities of our times.

Yet I am a very religious man and somewhat of a mystic. Certainly I have never been an atheist or even an agnostic. Broadly speaking, I do not believe that it is necessary to know everything, or indeed anything at all about God, and that there are plenty of other things to occupy our time and energies without worrying too much about God. I am unwilling to conceive of God; if I think about Him at all, it is not as a good friend who has a terrible job on His hands, if His entire concern is claimed by the human race. That any of us should hold such a belief—as that of the pre-occupation of God with our puny ambitions, desires, and miseries—argues presumption and egotism in the face of something so vast and terrifying as the universe—presumptuous, egotistical ants.

I am, however, deeply grateful toward whatever force has made this world full of children and dogs, and trees and streams and valleys, and the makings of good food and drink and love in all its manifestations. And I am grateful to Moses for having thought up the Ten Commandments, which is the best possible code for keeping the disorderly human race in order, and to Christ for giving us the Sermon on the Mount, which most of us might well study more carefully to our own advantage and that of our fellow men. But I am certainly not concerned with the arguments

between the divisions and sects of Christianity, from the Roman Catholic
Church down to the worst and most ignorant and insensible of the floor-
rolling splinter sects. And I have no interest in those tiresome theological
discussions over this or that bit of dogma which may be of interest to
dusty scholars and moldy priests, but can be of little interest or importance to the higher values which concern the long-suffering human race.

The present Pope is a very great man and statesman of fine character
and brilliant mind, but I doubt that he knows any more about a finite
God than you or I, or indeed the ranting Protestant preacher raving of
hell-fire and damnation, who must indeed actually offend God, wherever He is. These wild conceptions of the life after death, conceived and
born out of the idea of the vengeful Yahweh of the antique Jews and the
ignorance and illiteracy of the Middle Ages, have never had any appeal or
any reality for me. Least of all, have I ever had sympathy with those who
deny the God-given pleasures of this life, in the vague and unproven hope
of a better one; indeed, I cannot quite see how or why they should wish a
better life than this one, unless they themselves have muddled this one;
and, of course, it may be that they are merely the victims of a very dirty
trick indeed, having given up the pleasures of one existence for another
which is not there at all. Above all, I have always felt an instinctive contempt for all such beliefs, because they seem to corrupt and rot the very
dignity and stature which has been offered to man, and which he may
achieve if he chooses.

On the other hand, the long material and historical record of evil, cor-
ruption, intolerance, and bigotry of the Christian Church, both Roman
Catholic and Protestant, seemed to me, even as a boy with a passion for
reading, a very poor foundation or means of dissemination for the kindly, tolerant, ethical doctrines attributed to Christ. As I grew older, I was to
find in a great many churches, even in our own times, preachers and even
congregations who were sunk in ignorance, intolerance, bigotry, and just
plain foolishness.

So here I was, in my forties, seized and driven more and more by a
force which I did not understand, searching for some enlightenment
which might at least give form to a force of which I was aware but which
was utterly shapeless and formless because in all religious writing and
complex theological doctrine since the beginning of time, it had never
been clarified or formulated.

For a while I became at least part Deist and at times even approached being an animist worshiping "god" and the laws of the universe in the form of trees and rocks and springs. I was strongly aware of soils and trees and plants and animals, and my own very close association with them from the time I was a very small boy. It was only years later in the White Room of the Great House in Brazil that I began to discover what it was I sought, what it was that had given direction after a fashion to the whole of my life, and to find the end indeed of the chaos which had long since taken over a very busy and occupied and inquiring mind.

I had brought with me on the long trip from Ohio to Brazil a number of books which I had long wanted to read, and which I had not had the time; among them was a book called *Out of My Life and Thought* by Albert Schweitzer. Until then I had read only fragments of what Schweitzer had written. I knew that he had received the Nobel Peace Prize, but out of my worldly experience, I knew that it had become fashionable (and easy) for journalists to use such a figure as a subject for facile articles. At that time my own approach to Albert Schweitzer was a cynical one, for I knew by long experience the hysteria which, in a different way, is as typical of our times as it was of the Middle Ages. Radio, television, press agents, and many other elements made it possible regularly for the journalists and then the public to take up a more or less public figure anywhere, out of almost any world of human activity, and inflate him rapidly into a spectacular historical figure of world-shaking importance. Indeed, many of the famous people of our time have been made famous not through their own virtues, character, or accomplishments, but because the journalists and radio commentators were looking for someone who would make a "good story" that could be turned out quickly. The general public, but most of all the "literary" and semi-intellectual public, suffers twin afflictions in our times ... one of a close resemblance to a flock of sheep and the other of a kind of mass snobbery, which rises and subsides like the tides of the sea. And so I approached Schweitzer with crossed fingers and an attitude of doubt.

The earlier parts of the book are concerned almost entirely with the interpretation of Bach, the building and history of church organs, and considerable scholastic dissection into the question of whether Christ ever really existed or not. The subjects appealed to me in about that order. I love Bach's music. The author wrote in the most interesting fashion possible concerning organs and organ music, and because he was obviously a true scholar, he induced in me for the first time an interest in the dull, hairsplitting arguments of the theologians over the historical aspects of Christ's existence, a subject which has never seemed to me of first importance.

It was only when Schweitzer, trained as a theologian and preacher, arrived at the point where he felt that he must do more than merely preach and teach within the orbit of Christian doctrine, that the book took fire. Here was something I could understand ... a Christian who felt that he must do something tangible about being a Christian and decided that he could do most good and best satisfy his spirit by studying medicine and going into the work of actively and materially helping the more unfortunate members of the human race. It was at this point that for me the book began to have a real and significant meaning of great importance.

Here was no Lawrence of Arabia, an insignificant and psychopathic little man, trumped up by press agents and propaganda into a great world-shaking figure because he could be of use to a dying imperialistic power; this was not a Jean Paul Sartre, warming over for a public weakened and cowering after a disaster, and for the intellectual snobs, the negative and defeatist century-old philosophic doctrines of a Protestant Danish preacher. This was clearly a man, a robust man, who was a musician, a scientist, a philosopher, and perhaps above all a good man who asserted with vigor the essential and potential and *positive* dignity of Man. This was no super-specialist, engrossed entirely with what went on inside the narrow limits of his small, dark alley, nor yet the "pure" scientist detached, deformed, and isolated from the activities of his fellow men. This was a figure, so much needed in our narrow, frantic, overspecialized, materialist age, out of the Renaissance ... a full man, a complete man, exercising all his rights, developing all his talents, and with a vigorous, robust faith not in angels and harps or purgatories or hell-fire and damnation, but in the realities of Man himself. Above all, he was a Christian, and something better than the Christian as defined in and by our times.

As I read on in the pages Schweitzer had written, often painfully, and frequently when he was ill and exhausted from the labors of his day's work, the book grew more exciting and impressive. It seemed to me that here I found a man who had not only realized the talents and potentialities with which most men are endowed to some degree, but also a man whose philosophy was needed bitterly in this Age of Irritation and Frustration.

And then I came upon that passage which for me illuminated a whole new world of philosophy and religion and brought to me the clarification and the formalization of all the chaos which had been troubling me for years.

I will simply quote the passage in which Schweitzer created a phrase which provided me with the long-sought illumination. The scene is a barge on a river in the midst of Africa. It reads:

> Late on the third day, at the very moment when, at sunset, we were making our way through a herd of hippopotamuses, there flashed upon my mind, unforeseen and unsought, the phrase "Reverence for Life". The iron door had yielded; the path in the thicket had become visible. Now I had found my way to the idea in which affirmation of the world and of ethics are contained side by side. Now I knew that the ethical acceptance of the world and of life, together with the ideals of civilization contained in this concept, had a foundation in thought.

In this very great phrase, "Reverence for Life," I too found what I had sought for so long. It explained not only my thoughts, but the emotions and instincts which I had long experienced as a man. It is one of those phrases which, stimulating thought to an almost unbearable degree, illuminates the darkness like the switching on of a powerful light bulb—a phrase which is fecund and keeps breeding thought upon thought, conception upon conception upon conception. It was like the bursting of a rocket high in the darkness of the night air.

The phrase "Reverence for Life" brought the elements of the mechanistic and the material together with the ethical and even the spiritual, something that the church, in all its forms, has long since been unable to do, and the attempt at which it has apparently abandoned. It asserted the principle of affirmation rather than denial; it restored dignity to man. In a sense it defined God for the first time, for if God is not Life, He is merely a vaporous figment of the imagination and the delusion of the weak and the frightened, and of those very weak and fearful and ignorant who are so profoundly in need of help not from God so much as from their fellow men who are stronger, more intelligent, perhaps even more favored. In this phrase there was none of the sense of egotistical denial which has always imbued the Hindu faith and all its branches and has served as a blight upon the real progress toward the sun of most of the peoples it has influenced. In it there was none of the vengeance and bitterness of Judaism. In it there was the shadow of the glory of the Greeks.

To quote further: "Affirmation of life is the act by which man ceases to live unreflectively and begins to devote himself to his life with reverence in order to raise it to its true value. To affirm life is to deepen, to make more inward, and to exalt the will-to-live."

And still one more quotation which touched me very closely: "A Man is ethical only when life, as such, is sacred to him, that of plants and animals as well as that of his fellow men, and when he devotes himself helpfully to all life that is in need of help."

I began to understand what it was that had taken increasing possession of me for fifteen years, indeed for nearly a generation; why I had committed what some of my fellow men regarded as follies but which were not follies at all, why I had poured out time and energy and money in very large amounts for things on which there could not possibly be any material return. It explained to me why I have rarely said no to life and to experience, and why my sense of sin is notably deficient by the standards of most church dogma; why I have always enjoyed to the full the five senses which are a part of the glory of life, why there were certain shabby things, certain cheating and shortcutting, I was never able to practice without a sense of degradation and the violation of principles which I until now have never fully understood or put into words.

There are things which one does or one does not do, and in his heart even the savage knows what these are, and each time one violates this principle, the individual becomes more brutalized and sinks a little lower, not only in the esteem of others, but in the secret knowledge deep in his own heart—a knowledge which in time grows into a face which repels warmth and friendship and respect and makes of any man an outcast on sight from his fellow men, and above all, from discerning, intelligent, and good men.

Michelet knew this when he wrote, *"Dans la damnation le feu est le moindre chose: le suplice propre au damné est le progrès infini dans le vice et le crime, l'âme s'endurcissent, se dépravant toujours, s'enfonçant nécessairement dans le mal de la minute en progression géométrique pendant l'éternité."* (Being translated this means simply: "In damnation the fire is of little importance. The real punishment is the steady progression deeper and deeper into vice and crime, while the spirit grows steadily more calloused

and more depraved, lost constantly in the evil of the moment in a geometric progression into eternity.")

Even in the little business of a lie, the evil multiplies into more and
more lies as the good truth recedes farther and farther into the distance.
How much worse a hell is this than anything imagined by Dante,
although at times there are passages in the *Inferno* which come close to
such a conception. The real sins, the only sins, are those against Life itself
and the principle of Reverence for Life; and perhaps the worst sins of all
are those of intolerance and bigotry. Who can doubt this who has known
the poor shabby souls of the ignorant and the bigoted who believe they
alone know the path to Heaven, and would compel their fellow men by
persecution and force to follow that wretched, sterile, ungenerous path?
They live in a narrow hell of their own from which all beauty, all warmth,
all glory is forever forbidden them.

And so in this book when I have been writing about plants, animals,
and people, I now know why, and I know why my own life, which has
been a singularly fortunate one, has grown richer and more satisfactory
as instinctively and unconsciously I have moved toward a Reverence for
Life. This principle is known to every *good* farmer, as it is known to every
truly good and truly happy person. It is unfortunate that we are sometimes weak enough to sin against it, and each time we sin we suffer a
weakening of our dignity. The sin can be committed in countless ways,
from unkindness to the greedy ravaging of a forest, from the making of a
"smart deal" to the wasting of land which belongs not to us but to Life,
and which we hold only in trust for future generations.

In no other field of activity can the whole principle of the Reverence
for Life, which may indeed constitute the very basis of the preservation of
our civilization, be so thoroughly, easily, and profoundly understood and
exercised as in the field of agriculture, for, as I have pointed out many
times in this book, it is the only profession in which man deals constantly with *all* the laws of the universe and life. A productive agriculture of
high quality is the very foundation of our health, our vigor, even of our
intelligence; for intelligence expands, grows, and functions best in a
healthy and vigorous body. But agriculture is also one of the greatest of
civilizing influences, especially in those backward and ill-fed and confused nations which have become increasingly the gravest problem of the
world and of man himself. Often enough, it is more important in the

early stage of sociological and economic development and advance, than education in all but the most primitive sense.

Every good farmer practices, even though he may not understand clearly, the principle of Reverence for Life, and in this he is among the most fortunate of men, for he lives close enough to Life to hear the very pulsations of the heart, which are concealed from those whose lives are concentrated upon the unbalanced shabbiness of the completely material.